Solving Problems in Control

Other titles in the Series

Solving Problems in Vibrations, *J.S. Anderson and M. Bratos-Anderson*

Solving Problems in Fluid Mechanics Volume 1, *J.F. Douglas*

Solving Problems in Fluid Mechanics Volume 2, *J.F. Douglas*

Solving Problems in Soil Mechanics, Second Edition, *B.H.C. Sutton*

Solving Problems in Applied Thermodynamics and Energy Conversion, *G.J. Sharpe*

Solving Problems in Solid Mechanics Volume 1, *S.A. Urry and P.J. Turner*

Solving Problems in Solid Mechanics Volume 2, *S.A. Urry and P.J. Turner*

Solving Problems in Structures Volume 1, *P.C.L. Croxton and L.H. Martin*

Solving Problems in Structures Volume 2, *P.C.L. Croxton and L.H. Martin*

Solving Problems in Electrical Power and Power Electronics, *H.F.G. Gwyther*

Solving Problems in Surveying, *A. Bannister and R. Baker*

Solving Problems in Fluid Dynamics, *G.J. Sharpe*

Solving Problems in Control

R.J. Richards
University of Cambridge

Longman
Scientific &
Technical

Copublished in the United States with
John Wiley & Sons, Inc., New York

Longman Scientific & Technical,
Longman Group UK Limited,
Longman House, Burnt Mill, Harlow,
Essex CM20 2JE, England
and Associated Companies throughout the world.

*Copublished in the United States with
John Wiley & Sons, Inc., 605 Third Avenue, New York, NY 10158*

© Longman Group UK Ltd 1993
First published 1993

British Library Cataloguing in Publication Data
A catalogue entry for this title is available from the British Library.
ISBN 0582 03298 9

Library of Congress Cataloging-in-Publication Data
A catalogue entry for this title is available from the Library of Congress.
ISBN 0470 22076 7 (USA only)

Set by 4 in 10/12 Compugraphic Times
Printed in Malaysia by PA

Contents

Preface   vii

1 Fundamental material 1
   Introduction 1
   Differential equations 2
   Laplace transforms 6
   Convolution integral 13
   Transfer functions 15
   Problems 17

2 Linear system modelling 22
   Introduction 22
   System inputs 23
   Electrical systems 24
   Mechanical systems 29
   Electromechanical systems 37
   Hydraulic systems 39
   Process systems 40
   Problems 49

3 System representation by diagrams 53
   Benefits 53
   Signal flow graphs 53
   Block diagrams 58
   Problems 65

4 Systems with simple feedback 70
   Feedback 70
   Problems 78

5 System poles, zeros and stability 81
   Rational transfer function 81
   System poles 81
   Stability 82
   System zeros 86
   Problems 89

6 Frequency response 93
       Frequency transfer function 93
       Phasor diagrams 94
       Polar plots 101
       Inverse polar plots 107
       Logarithmic plots 113
       Problems 122

7 Additional controller terms 128
       Integral action 128
       Derivative action 130
       PID control 132
       Velocity feedback 136
       Feedforward terms 141
       Cascade control 146
       Problems 149

8 Stability and the time domain 154
       Stability of closed loop systems 154
       Routh−Hurwitz criterion 154
       Root locus 156
       Problems 169

9 Stability and the frequency domain 173
       The Nyquist criterion 173
       Relative stability 181
       Stability and the inverse Nyquist plot 184
       Stability and the Bode plot 188
       Closed loop frequency response 192
       M and N circles and the Nichols chart 194
       Nichols chart 196
       Problems 199

10 Phase compensators and other controllers 204
       Phase compensators 204
       Feedforward control 222
       Cascade control 222
       Loop interaction 222
       Problems 226

Index 231

Preface

This book aims to help students by illustrating, by means of a series of worked examples, the fundamental steps underlying an understanding of control principles. Attention is restricted to the important basis which comes from continuous linear single-input single-output systems. The approach is progressive starting with the mathematical methods used most frequently in the study of single-input single-output systems. Because much of the teaching of control is necessarily done in terms of a mathematical model of a device or plant, examples of the formation of deterministic models of simple physical systems follow. Manipulation of these models leads to the introduction of control and the use of both time domain and frequency response methods. In particular the assessment of the stability of closed loop negative feedback control systems forms the central feature of this.

Such a text as this cannot be exhaustive either in terms of material covered or in its range of examples. It is hoped that in its limited length and fundamental context it gives wide accessibility to students of many engineering disciplines and assists them to overcome the early stumbling blocks of representing physical systems by mathematical equations and the subsequent manipulation of these in a dynamics and control context. The further development of the subject relies on a firm foundation of this material.

1

Fundamental material

Introduction

Within the context of control, we are dealing with controlling or influencing the behaviour of a chosen system, be it mechanical, electrical, or chemical, or some combination of these, by changing an input to that system according to a chosen rule. In order to develop a suitable control rule for a specific system one may start by having a knowledge of the behaviour of the physical system, being able to represent that knowledge in a mathematical expression and then using that expression or 'model' as a basis for control studies. The further stages of implementation and tuning of the controller in use then follow.

In this text the use of modelling and methods of fundamental control studies are reviewed briefly and illustrated by a series of examples. In particular, systems with single input and single output continuous deterministic relationships which are modelled by linear differential equations are dealt with. The field of control has been extended by the use of many new techniques, e.g. the use of multi-input multi-output systems, identification in system modelling, and computer aided design packages. The basis of these lies in the material covered here and the plan in the following chapters is to take the reader through the basics of control by means of the examples. As in most areas of study the possibilities of systems and examples are almost boundless and the actual approach to any one problem frequently contains alternatives. Such a text cannot therefore be exhaustive but it is intended that illustrations are given which help specifically in seeing how the fundamentals of simple system dynamics and control are used and developed.

The representation of physical behaviour by a mathematical form is called a mathematical 'model' of the system. The way in which the system changes with time, normally detected by measuring an output or product quality, is expressed in a system model as a differential equation. The system reacts to an independent input variable to give an output. Usually in control our major concern is with how *changes* in inputs affect the output and how well the output responds to changes in the demand input.

If the input is a defined function of time, such as a step change in value, then the differential equation expressing the change in output becomes a function of time, i.e. has time as the independent variable. Within the study of single-input single-output continuous control systems it has become common practice to solve such equations by the method of Laplace transforms. The use of these transforms in turn leads to a description of the system dynamics using the 'transfer function'. This first chapter starts with a consideration of the form and solution of linear differential equations such as will be found to represent simple system behaviour and shows by examples how these lead to the transfer function and solution by this route.

In the second chapter experience in developing the models and solutions for a variety of low order systems is obtained. Such system representation is usually made clearer by the use of block diagrams or signal flow diagrams and the use of these in simplifying the algebra and overall input to output relationships is covered in Chapter 3. The material to this point is in terms of general single-input—single-output systems. (It is the subject content of these chapters, the overall introduction to general system representation, that frequently forms an initial major difficulty in control studies. A firm understanding here prepares the ground for the introduction of the control aspects.) However, feedback occurs naturally in some systems or is readily added. Initial examples of feedback are introduced in Chapter 4. The key property of control systems is that of stability, the bounded responses of the system to an input. This property of a system model is related to the model's differential equation, in particular to the roots of the denominator of the system's transfer function. These roots, the system 'poles', together with the 'zeros', the roots of the transfer function's numerator, and stability are investigated by examples in Chapter 5.

Of particular use in the analysis of system performance and controller synthesis is the response of a system to a forcing sinusoidal input. This may be handled in a number of ways, e.g. by phasor diagrams or the more useful methods in the control field of polar (Nyquist) plots and logarithmic (Bode) plots. Examples of the study of frequency response using these techniques, in both open loop and feedback systems, are given in Chapter 6. The remaining chapters develop the feedback control first introduced in Chapter 4. In Chapter 7 the simple controller is developed through the addition of integral, derivative and velocity feedback terms. Examples of feedforward and combined (cascade) controllers are also included here. The study of stability through the 'time domain' is followed up in Chapter 8 with examples of the use of the Routh—Hurwitz stability criterion and root locus plots. Stability remains the key topic in Chapter 9 which returns to the 'frequency domain'. In the final chapter the use of additional phase compensators forms the major part, using principally the frequency domain methods. Chapter 10 also includes the design of simple compensators to suppress the effects of secondary inputs, i.e. disturbances and noise in the system.

Differential equations

Although the overall relationship is initially between the input to the system and the output, the input itself is a function of time and hence it is possible to reduce the equation to one relating the output to the variable 'time'. The process of arriving at the input/output relationship is called 'modelling' and is covered in Chapter 2. If the coefficients in the differential equation are constant then the system is called time invariant. If the system has the property of superposition, i.e. that the output resulting from a sum of inputs is equal to the sum of outputs resulting from each of those inputs applied separately, then it is also said to be linear. The class of system which can be represented by ordinary differential equations having these properties, and which can be handled by a set of standard procedures, is the 'linear time-invariant continuous system'. Unless otherwise stated it is assumed that the system being considered is of this form.

1.1 Tank flow model

Liquid flows at a variable rate $f(t)$ into a tank and leaves through an open valve at the bottom. The rate of flow out through this valve is proportional to the depth of liquid $h(t)$ in the tank. The resulting relationship between the depth of liquid $h(t)$ m in the tank and the variable flow $f(t)$ m^3 s^{-1} into the tank is given by the first order differential equation

$$3 \frac{dh}{dt} + 0.1 \, h = f(t)$$

(i) If the input into the empty tank is a sudden addition of 1.5 m^3 which may be considered to be instantaneous, express the resulting changes in height as a function of time.

Show that this is equivalent to opening the outlet valve when the initial height in the tank is 0.5 m and there is no additional input.

(ii) If the input flow $f(t)$ is maintained constant at a constant flow rate of 0.01 m^3 s^{-1} into the tank, which again starts empty, plot the change in level of the liquid in the tank with time.

Solution (i) The tank is initially empty and there is thus the initial boundary condition with $h = 0$. An addition of liquid assumed to take place within a very short time will therefore raise the level in the tank, the flow in from this impulse of liquid then stops and the resulting change in liquid level is from this new 'initial' condition. Thus the flow from the tank, and the change in liquid level, is the same as starting from an initial level and opening the outlet valve.

The response of a first order system, $dh/dt + h/T = F(t)$, to an impulse of magnitude A is

$$h = A \, e^{-t/T}$$

Thus writing the system equation in this form,

$$\frac{dh}{dt} + \frac{h}{30} = \frac{f(t)}{3}$$

$T = 30$ (the system time constant) and $A = 1.5/3$ so that

$$h = 0.5 \, e^{-t/30}$$

(ii) For the step input $f(t) = 0.01 \text{ m}^3 \text{ s}^{-1}$ and initial condition $h(0) = 0$, we may solve using the complementary function plus particular integral. For

$$\frac{dh}{dt} + \frac{h}{30} = \frac{0.01}{3}$$

the complementary function is

$$h = C\, e^{-t/30}$$

and the particular integral is

$$h = 0.1$$

The general solution is

$$h = C\, e^{-t/30} + 0.1$$

Putting in the boundary condition $h = 0$ at $t = 0$ gives $C = -0.1$ and

$$h = 0.1(1 - e^{-t/30})$$

The time response showing growth in the liquid level is given in Fig. 1.1.

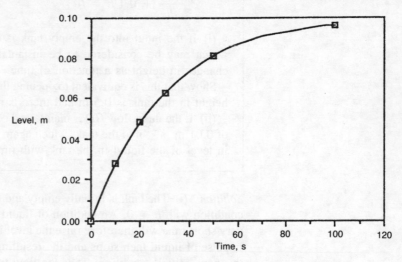

Figure 1.1 Step response of first order system

1.2 Response of second order system to a step input

A particular linear system model is given by the second order ordinary differential equation

$$\frac{d^2y}{dt^2} + 3\,\frac{dy}{dt} + 2y = u(t)$$

For $t < 0$, y and its time derivative and $u(t)$ are zero.

(i) Find the response $y(t)$ to a unit step input, $\mathcal{H}(t)$, i.e. $u(t) = 1$ for $t \geq 0$.

(ii) Based on this solution determine the response of the same system to a unit magnitude pulse of duration 2 s.

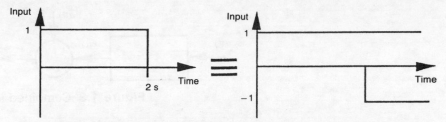

Figure 1.2 Pulse input as a summation of two steps

Solution (i) Given the system equation the solution, i.e. the way in which the system represented by this equation is seen to behave, may be found by the standard complementary function plus particular integral method.

The complementary function from

$$\frac{d^2y}{dt^2} + \frac{3dy}{dt} + 2y = 0$$

yields

$$y = A\,e^{-2t} + B\,e^{-t}$$

The particular integral for $u(t) = 1$, $t \ge 0$, is

$$y = 0.5$$

The general solution is

$$y = A\,e^{-2t} + B\,e^{-t} + 0.5$$

Substituting for the boundary conditions, this solution becomes

$$y = 0.5\,e^{-2t} - e^{-t} + 0.5$$

(ii) The pulse of duration 2 s is treated as two consecutive steps but of opposite sign, Fig. 1.2.

The solution is thus the sum of the solutions resulting from each of these inputs if applied separately with a delay between them of 2 s, i.e.

$$y = 0.5\,e^{-2t} - e^{-t} + 0.5$$
$$- \mathcal{H}(t-2)(0.5\,e^{-2(t-2)} - e^{-(t-2)} + 0.5)$$

where $\mathcal{H}(t-2) = 0$ for $t<2$, 1 for $t \ge 2$.

1.3 Systems in series

The input $u(t)$ is applied to a first order system and gives a response $x(t)$. This forms the input to a second system which is also subject to a further input $d(t)$ at the same point. The final output is $y(t)$. If the full system starts from zero initial conditions, and both $u(t)$ and $d(t)$ are unit steps from $t = 0$, plot the resulting output $y(t)$ as it evolves with time.

The defining equations for the two parts of the system are

$$\frac{dx}{dt} + 3x = u \quad \text{and} \quad \frac{dy}{dt} + 2y = d + x$$

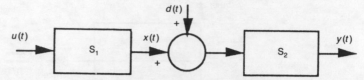

Figure 1.3 Combined inputs

Solution　The problem is made clearer, as in nearly all cases, by a diagram, Fig. 1.3.

Using the system equation first solve for S_1 with a unit step input to give

$$x = \tfrac{1}{3}(1 - e^{-3t})$$

For the second system the input is now a unit step, $d(t)$, plus $x(t)$. We may solve for the inputs separately or combine them immediately in the formation of the solution so that the defining equation for the output $y(t)$ is

$$\frac{dy}{dt} + 2y = \tfrac{4}{3} - \tfrac{1}{3}e^{-3t}$$

yielding the solution, Fig. 1.4,

$$y = \tfrac{2}{3} - e^{-2t} + \tfrac{1}{3}e^{-3t}$$

Note that the superposition principle applies for the linear system S_2 and the combination of two first order systems in series gives a second order system.

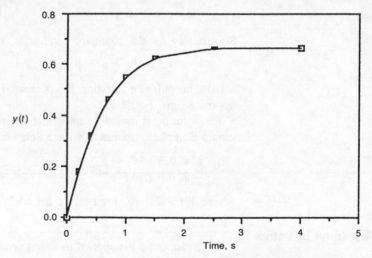

Figure 1.4 Response to summed inputs

Laplace transforms

Let $f(t)$ be a function of time t, with $f(t) = 0$ at $t < 0$. The Laplace transform of $f(t)$, denoted $F(s)$ or $f(s)$, is defined by the relationship

$$F(s) = \int_0^\infty f(t)\, e^{-st} dt$$

where s is the complex variable, $\sigma+j\omega$, and the function exists only for $0 \le t < \infty$. A benefit aiding its use in the context of linear constant coefficient differential equations, arising in fundamental dynamic and control studies, is the ability to establish a table of transform pairs, Table 1.1, and manipulation theorems, Table 1.2. These lead to rapid solutions in standard forms. If a differential equation relates an 'input' to an 'output' of a system then the use of this transform similarly relates the transformed variables. Under specific conditions this leads to the concept of the transfer function.

Table 1.1 Laplace transform pairs (from Richards RJ 1979 *An Introduction to Dynamics and Control*, Harlow, Longman pp. 10−11)

$f(t)$	$F(s)$
Unit impulse, $\delta(t)$	1
Unit step, $1(t)$	$1/s$
t	$1/s^2$
t^2	$2/s^3$
$t^n\,(n = 1, 2, 3, \ldots)$	$n!/s^{n+1}$
$\dfrac{df(t)}{dt}$	$sF(s) - f(0)$
$\dfrac{d^n f(t)}{dt^n}$	$s^n F(s) - s^{n-1} f(0) - s^{n-2} \dot{f}(0) \ldots f^{n-1}(0)$
e^{-at}	$1/(s+a)$
te^{-at}	$1/(s+a)^2$
$t^n e^{-at}$	$n!/(s+a)^{n+1}$
$\dfrac{1}{(b-a)} \cdot (e^{-at} - e^{-bt})$	$\dfrac{1}{(s+a)(s+b)}$
$\dfrac{1}{(b-a)} \cdot (be^{-bt} - ae^{-at})$	$\dfrac{s}{(s+a)(s+b)}$
$\dfrac{1}{ab} \left[1 + \dfrac{1}{(a-b)} \cdot (be^{-at} - ae^{-bt}) \right]$	$\dfrac{1}{s(s+a)(s+b)}$
$\sin \omega t$	$\dfrac{\omega}{(s^2 + \omega^2)}$
$\cos \omega t$	$\dfrac{s}{(s^2 + \omega^2)}$
$\sinh \omega t$	$\dfrac{\omega}{s^2 - \omega^2}$
$\cosh \omega t$	$\dfrac{s}{s^2 - \omega^2}$

Table 1.1

$f(t)$	$F(s)$
$e^{-at} \sin \omega t$	$\dfrac{\omega}{(s+a)^2+\omega^2}$
$e^{-at} \cos \omega t$	$\dfrac{s+a}{(s+a)^2+\omega^2}$
$\dfrac{\omega_n}{\sqrt{(1-\zeta^2)}} \cdot e^{-\zeta\omega_n t} \sin[\omega_n\sqrt{(1-\zeta^2)}t]$	$\dfrac{\omega_n^2}{s^2+2\zeta\omega_n s+\omega_n^2}$
$\dfrac{-1}{\sqrt{(1-\zeta^2)}} \cdot e^{-\zeta\omega_n t} \sin[\omega_n\sqrt{(1-\zeta^2)}t-\phi]$ $\text{where } \phi = \tan^{-1}\dfrac{\sqrt{(1-\zeta^2)}}{\zeta}$	$\dfrac{s}{s^2+2\zeta\omega_n s+\omega_n^2}$
$1 - \dfrac{1}{\sqrt{(1-\zeta^2)}} \cdot e^{-\zeta\omega_n t} \sin[\omega_n\sqrt{(1-\zeta^2)}t+\phi]$ $\text{where } \phi = \tan^{-1}\dfrac{\sqrt{(1-\zeta^2)}}{\zeta}$	$\dfrac{\omega_n^2}{s(s^2+2\zeta\omega_n s+\omega_n^2)}$

1.4 Tank flow problem with Laplace transforms

Consider again the differential equation which models the tank system of Example 1.1, i.e.

$$3\frac{dh}{dt} + 0.1h = f(t)$$

(i) Use the table of Laplace transforms to form this relationship in terms of the transformed variables.

(ii) Hence derive the solution for the change of height of liquid, h, with time when the input to the system is an impulse of magnitude 2 m^3 followed immediately by constant flow of 0.01 m^3 s^{-1}.

Solution (i) From the table of transforms the transform of the time derivative is $sh(s)-h(0)$ so that the differential equation relationship becomes

$$3sh(s) - 3h(0) + 0.1h(s) = f(s)$$

or

$$h(s) = \frac{f(s)}{3s+0.1} + \frac{3h(0)}{3s+0.1}$$

(ii) For the combined impulse plus step input of liquid,

$$f(s) = 2 + \frac{0.01}{s}$$

Table 1.2 Laplace transform theorems (from Richards RJ 1979 *An Introduction to Dynamics and Control*, Harlow, Longman pp. 11–12)

1. *Linearity*

   $$\mathcal{L}[f_1(t) + f_1(t)] = \mathcal{L}[f_1(t)] + \mathcal{L}[f_2(t)]$$

2. *Constant multiplication*

   $$\mathcal{L}[af(t)] = a\mathcal{L}[f(t)]$$

3. *Differentiation*

   $$\mathcal{L}\left[\frac{df(t)}{dt}\right] = s\mathcal{L}[f(t)] - f(0)$$

   $$\mathcal{L}\left[\frac{d^2f(t)}{dt^2}\right] = s^2\,\mathcal{L}[f(t)] - sf(0) - \dot{f}(0)$$

4. *Integration*

   $$\mathcal{L}\left[\int f(t)dt\right] = \frac{F(s)}{s} + \frac{[\int f(t)dt]_{t=0}}{s}$$

5. *Translated function*

   $$\mathcal{L}[f(t-\alpha)] = e^{-\alpha s}\,\mathcal{L}[f(t)] \text{ with } f(t-\alpha) = 0 \text{ for } t < \alpha$$

6. *Multiplication of $f(t)$ by $e^{-\alpha t}$*

   $$\mathcal{L}[e^{-\alpha t}f(t)] = F(s+\alpha)$$

7. *Time scale change*

   $$\mathcal{L}\left[f\left(\frac{t}{a}\right)\right] = aF(as)$$

8. *Final value theorem*

   $$\lim_{t\to\infty} f(t) = \lim_{s\to 0} sF(s)$$

9. *Initial value theorem*

   $$\lim_{t\to 0} f(t) = \lim_{s\to\infty} sF(s)$$

10. *Heaviside expansion formula*: For the rational algebraic function $F(s) = A(s)/B(s)$

    $$f(t) = \mathcal{L}^{-1}[F(s)] = \sum_{i=1}^{n} \frac{A(s_i)}{B'(s_i)}\, e^{s_i t}$$

    where s_i are the n distinct roots of the characteristic equation $B(s) = 0$ and B' signifies differentiation with respect to s

11. *Convolution integral*: The Laplace transform of the convolution integral is given by

    $$\mathcal{L}\left[\int_0^t f_1(t-\tau)f_2(\tau)d\tau\right] = F_1(s)F_2(s)$$

and with the zero initial depth

$$h(s) = \frac{2}{3s+0.1} + \frac{0.01}{s(3s+0.1)}$$

$$= \frac{2}{3}\left(\frac{1}{s+0.0333}\right) + \frac{0.01}{0.1}\left(\frac{1}{s} - \frac{1}{s+0.0333}\right)$$

to give, on term by term inversion,

$$h(t) = \frac{2}{3}\,e^{-0.0333t} + \frac{1}{10}\,(1-e^{-0.0333t})$$

The full response is the sum of the responses to the impulse and step evaluated separately. A check on the solution shows that $h(0) = 0$ and $h(\infty) = 0.1$, as the head in the tank adjusts to balance input and output flows. Note also the use of partial fractions with the Laplace transforms to obtain invertable forms.

1.5 System response to a ramp input

(i) A simplified representation of a system's dynamic behaviour is given by the first order equation

$$\frac{dx}{dt} + 2x = u(t)$$

What is the response of the output with time if the input has the ramp form $u(t) = 3t$ for $t \geq 0$ and $x(0) = 0$?

(ii) Additions to the system lead to a defining equation

$$\frac{d^2x}{dt^2} + 5\frac{dx}{dt} + 6x = u(t)$$

What is now the response of the system to both the same ramp and a unit step if (a) the system is at initial zero conditions and (b) the initial conditions are $x(0) = 0.1$ m, $dx/dt(0) = 0.1$ m s^{-1}?

Solution (i) Substituting for $u(t)$ and taking Laplace transforms leads to

$$sx(s) + 2x(s) = \frac{3}{s^2}$$

i.e.

$$x(s) = \frac{3}{s^2(s+2)}$$

Division into partial fractions and evaluation of each coefficient (e.g. by the 'cover-up rule') gives

$$x(s) = \frac{1.5}{s^2} - \frac{0.75}{s} + \frac{0.75}{s+2}$$

which on inversion gives the solution

$$x(t) = 1.5t - 0.75 + 0.75\ e^{-2t}$$

(ii) For the extended second order system and, (a) the same ramp input

$$s^2x(s) + 5sx(s) + 6x(s) = \frac{3}{s^2}$$

Rearranging and expanding into partial fractions

$$x(s) = \frac{3}{s^2(s+2)(s+3)}$$

$$= 3\left(\frac{1}{6}\frac{1}{s^2} - \frac{5}{36}\frac{1}{s} + \frac{9}{36}\frac{1}{s+2} - \frac{4}{36}\frac{1}{s+3}\right)$$

to give

$$x(t) = 0.5t - 0.417 + 0.75\ e^{-2t} - 0.333\ e^{-3t}$$

Where, (b), the input to this system is the unit step,

$$x(s) = \frac{1}{s(s+2)(s+3)}$$

leading to partial fractions

$$x(s) = \frac{1}{6}\frac{1}{s} - \frac{1}{2}\frac{1}{s+2} + \frac{1}{3}\frac{1}{s+3}$$

and to the solution

$$x(t) = 0.1667 - 0.5\ e^{-2t} + 0.333\ e^{-3t}$$

Note that this step response could be obtained from the ramp response by dividing by three [because the $u(t) = 3t$] and then differentiating with respect to time.

If the initial conditions are not 'zero' the system will have an 'unforced' component which will be additional to the responses evaluated above. This is evaluated with $u(t) = 0$. Then, on taking Laplace transforms,

$$s^2x(s) - 0.1s - 0.1 + 5sx(s) - 5 \times 0.1 + 6x(s) = 0$$

Rearranging to give

$$x(s) = \frac{0.6}{s^2+5s+6} + \frac{0.1s}{s^2+5s+6}$$

requires further inversion of the transformed terms to obtain the time solution. This can be done using the above procedures or we can note that the first term is what we would get from an impulse input (the derivative of a step) and the second term is what we would get from the further derivative, shown by the s in the numerator. The *additional* terms to take account of the imposed initial conditions are thus

$$x(t) = 0.6(e^{-2t} - e^{-3t}) + 0.1(-2 e^{-2t} + 3 e^{-3t})$$
$$= 0.4 e^{-2t} - 0.3 e^{-3t}$$

It is readily checked that this solution satisfies the initial conditions.

1.6 Solution of fourth order system

A fourth order system subject to a unit step input is represented by the equation

$$x'''' + 7x''' + 18x'' + 22x' + 12x = 1$$

with ′ representing the differential with respect to time t. What is the solution of this equation given all-zero initial conditions?

Solution Take Laplace transforms with the zero initial conditions for this fourth order equation,

$$(s^4 + 7s^3 + 18s^2 + 22s + 12)x(s) = \frac{1}{s}$$

to give

$$x(s) = \frac{1}{s(s^4 + 7s^3 + 18s^2 + 22s + 12)}$$

$$= \frac{1}{s(s+2)(s+3)(s^2 + 2s + 2)}$$

This yields, on putting it into partial fractions,

$$x(s) = \frac{1}{12}\frac{1}{s} - \frac{1}{4}\frac{1}{(s+2)} + \frac{1}{15}\frac{1}{(s+3)} + \frac{1}{10}\frac{(s-1)}{(s^2 + 2s + 2)}$$

The last of these terms, the quadratic factor, may be more readily inverted if expressed as

$$\frac{s+1}{(s+1)^2 + 1} - \frac{2}{(s+1)^2 + 1}$$

to give the final solution

$$x(t) = 0.0833 - 0.25 e^{-2t} + 0.0667 e^{-3t}$$
$$+ 0.1 e^{-t}\cos t - 0.2 e^{-t}\sin t$$

Now that a quick (but not infallible) check can be made by putting $t = 0$ in this solution. This gives $x(0) = 0$ as required.

1.7 Final value theorem

Use the final value theorem to determine the value of y as $t \to \infty$ where $y(t)$ is given in turn by each of the equations,

(i) $\dfrac{dy}{dt} + y = 1$ and

(ii) $$\frac{d^2y}{dt^2} + \frac{dy}{dt} = 1$$

Solution   (i) Establish the transformed equation

$$(s+1)y(s) = \frac{1}{s}$$

Then

$$y(s) = \frac{1}{s(s+1)}$$

and use of the final value theorem leads to

$$y(t)_{t \to \infty} = \lim_{s \to 0} \left[\frac{s}{s(s+1)}\right] = 1$$

(ii) For the second case

$$y(s) = \frac{1}{s^2(s+1)}$$

and

$$y(t)_{t \to \infty} = \lim_{s \to 0} \left[\frac{s}{s^2(s+1)}\right] = \infty$$

Thus the output of this system continues to grow in the presence of a steady input. It is the equivalent of the first system subjected to a unit ramp input [i.e. $u(t) = t$].

Convolution integral

The convolution operation relates the general input to a system, $u(t)$, via the response of the system, $g(t)$, to a unit impulse, $\delta(t)$, to the output, $x(t)$. The impulse function response $g(t)$ is also known as the weighting function and is a property of the system alone for linear systems. For the input $u(t)$ the system response is the 'convolution integral' of the function $g(t)$ and the input $u(t)$:

$$x(t) = g(t)*u(t)$$
$$= \int_0^t g(t-\tau)u(\tau)d\tau$$
$$= \int_0^t g(\tau)u(t-\tau)d\tau$$

The Laplace transformed equivalent is

$$x(s) = G(s)u(s)$$

with $G(s)$ the transform of the impulse response function $g(t)$.

The response of a first order system to a unit impulse input is given by

$$g(t) = e^{-at}$$

(i) Using the convolution equation directly determine the response to (a) a unit step and (b) a ramp $u(t) = bt$. Check the expressions using the Laplace transformed equivalent.

(ii) If the step input is discontinued after a time T what is the time response now? Check by use of the Laplace transform solution.

How are these solutions used for higher order system responses?

Solution (i) Direct substitution into the convolution equation with the *step* input $u(t) = 1$ gives

$$x(t) = \int_0^t e^{-a(t-\tau)}.1.d\tau$$

$$= e^{-at} \int_0^t e^{a\tau}d\tau = \frac{e^{-at}}{a}(e^{at}-1)$$

$$= \frac{1-e^{-at}}{a}$$

Direct use of the Laplace transform gives

$$x(s) = \frac{1}{(s+a)}\frac{1}{s} = \frac{1}{a}\left(\frac{1}{s} - \frac{1}{s+a}\right)$$

and

$$x(t) = \frac{1}{a}(1-e^{-at}) \text{ as before}$$

For the *ramp* input $u(t) = bt$ and the convolution integral now gives

$$x(t) = \int_0^t e^{-a(t-\tau)}.b\tau.d\tau$$

$$= b\, e^{-at} \int_0^t e^{a\tau}\tau d\tau$$

Evaluation of the integral of the product and insertion of the limits gives in turn

$$x(t) = b\, e^{-at}\left(\frac{t\, e^{at}}{a} - \frac{e^{at}}{a^2} - 0 + \frac{1}{a^2}\right)$$

$$= \frac{b}{a}\left[t - \frac{1}{a}(1 - e^{-at})\right]$$

Direct use of the Laplace transforms yields

$$x(s) = \frac{1}{(s+a)}\frac{b}{s^2} = b\left(\frac{1/a}{s^2} - \frac{1/a^2}{s} + \frac{1/a^2}{s+a}\right)$$

to give as before

$$x(t) = \frac{b}{a}\left[t - \frac{1}{a}(1 - e^{-at})\right]$$

(ii) If in the case of the step input the input is discontinued at $t = T$ this is equivalent to the initial step continuing but a second equal magnitude step but of opposite sign being added at this time T. Thus the output will be

$$x(t) = \frac{1}{a} (1 - e^{-at}) - \mathcal{H}(t - T). \frac{1}{a} [1 - e^{-a(t-T)}]$$

with $\mathcal{H}(t - T)$ having its usual meaning. Direct use of the Laplace transform where now

$$x(s) = \frac{1}{s+a} \left(\frac{1}{s} - \frac{e^{-Ts}}{s} \right)$$

leads to the same two terms on inversion. For higher order systems partial fractions give rise to terms like those appearing in first and second order systems. In general the use of the transformed variables will achieve a solution more readily than the convolution integral used from its definition.

Transfer functions

In the general transformation from the original differential equation using Laplace transforms, it is seen that the 'initial conditions' appear in the transformed equation. If these initial conditions are zero, or the dynamics can be re-expressed in a form which gives variables having zero initial values, then a direct ratio of the transformed 'input' and 'output' variables can be formed. The 'transfer function', $G(s)$, of the system is the ratio of the Laplace transform of the output, $x(t)$, to the Laplace transform of the input, $u(t)$, given zero initial conditions. It is also the transform of the impulse response function $g(t)$,

$$G(s) = \frac{X(s)}{U(s)} \left[\text{or} \frac{x(s)}{u(s)} \right]$$

Variables may be defined explicitly in terms of deviations from a 'steady state' so that initial values are zero.

1.9 Second order transfer function and response to step and ramp

A second order mechanical system is driven by a controller signal $u(t)$ (without feedback), such that the displacement $x(t)$ is given by

$$\frac{d^2x}{dt^2} + a \frac{dx}{dt} + bx = u(t)$$

(i) Express this in terms of the transformed variables $x(s)$ and $u(s)$ for general initial conditions and then form the transfer function relating $x(s)$ to $u(s)$.
(ii) If $a = 3$, $b = 2$ and $u(t) = c + kt$ evaluate $x(t)$.

Solution (i) Transforming the expression we have

$$(s^2 + as + b)x(s) - sx(0) - x'(0) - ax(0) = u(s)$$

so that

$$x(s) = \frac{u(s)}{s^2 + as + b} + \frac{x'(0) + (s+a)x(0)}{s^2 + as + b}$$

The transfer function between $u(s)$ and $x(s)$ is, from the first term,

$$G(s) = \frac{x(s)}{u(s)} = \frac{1}{s^2 + as + b}$$

(ii) For the specified input

$$u(s) = \frac{c}{s} + \frac{k}{s^2}$$

to give, with $a = 3$ and $b = 2$,

$$x(s) = \frac{c}{s(s+1)(s+2)} + \frac{k}{s^2(s+1)(s+2)}$$

Taking the second of these terms, for the ramp input, and expressing it in partial fractions,

$$x(s)_{ramp} = k\left(\frac{0.5}{s^2} - \frac{0.75}{s} + \frac{1}{s+1} - \frac{0.25}{s+2} \right)$$

i.e.

$$x(t)_{ramp} = k(0.5t - 0.75 + e^{-t} - 0.25\, e^{-2t})$$

With suitable gain adjustment, the part of the response coming from the step input may be obtained by differentiating the output from the ramp input. That is,

$$x(t)_{step} = \frac{c}{k}\, (0.5 - e^{-t} + 0.5\, e^{-2t})$$

The full response, by superposition for the linear system, is

$$x(t) = x(t)_{ramp} + x(t)_{step}$$

1.10 Transfer function after linearization

The input/output relationship of a non-linear system is given by the non-linear differential equation

$$\frac{d^2x}{dt^2} + \frac{dx}{dt} + x^2 = u(t)$$

Linearize this equation about the equilibrium point $x_0, u_0 = (0, 0)$ and derive the appropriate transfer function. Hence derive the response of the system to small changes in $u(t)$.

Solution As a means of establishing the equation for deviations from the set equilibrium conditions write the equation for $x_0 + \delta x$, $u_0 + \delta u$. Then

$$\frac{d^2(x_0+\delta x)}{dt^2} + \frac{d(x_0+\delta x)}{dt} + (x_0 + \delta x)^2 = (u_0 + \delta u)$$

Subtraction of the equilibrium condition, and for x_0, u_0 zero, and omitting the higher order term δx^2 the relationship becomes

$$\frac{d^2\delta x}{dt^2} + \frac{d\delta x}{dt} = \delta u$$

The transfer function for the changes in output will be

$$\frac{\delta x(s)}{\delta u(s)} = \frac{1}{s(s+1)}$$

The response to a small impulse, keeping the system in the linear approximation region, will be of the form $k(1-e^{-t})$. For a sustained input such as a step the output will continue to grow because of the integrator term $(1/s)$ in the transfer function and the region in which the approximation is valid will be exceeded.

Problems

1 Solve the following differential equations to give x as a function of time, $x(t)$. Use methods other than the Laplace transform.

(i) $\dfrac{dx}{dt} + 3x = u(t)$, $u(t) = 0$ for $t<0$, 3 for $t\geq0$
                                            $x(0) = 0$

(ii) $\dfrac{d^2x}{dt^2} + 5\dfrac{dx}{dt} + 6x = u(t)$, $u(t) = 0$ for $t<0$, 3 for $t\geq0$

           $x(0) = 0$, $\dfrac{dx}{dt}(t=0) = 0$

(iii) $\dfrac{d^2x}{dt^2} + \dfrac{dx}{dt} + 2x = u(t)$, $u(t) = 0$ for $t<0$, 3 for $t\geq0$

           $x(0) = 0$, $\dfrac{dx}{dt}(t=0) = 0$

In each case use the form of the solution to sketch the response $x(t)$ to the given $u(t)$ and evaluate $x(t=2)$. Derive also the response in the variable x if the input u is a unit impulse, $\delta(t)$.

Answer
(i) $x(t) = 1-e^{-3t}$, 0.9975, e^{-3t}
(ii) $x(t) = e^{-3t}-1.5\,e^{-2t}+0.5$, 0.4750, $-e^{-3t}+e^{-2t}$
(iii) $x(t) = 1.5[1-2\sqrt{\tfrac{2}{7}}\,e^{-0.5t}\sin(\tfrac{\sqrt{7}}{2}t+\tan^{-1}\sqrt{7})]$, 1.886, $\tfrac{2}{\sqrt{7}}\,e^{-0.5t}$
     $\sin(\tfrac{\sqrt{7}}{2}t)$

2 A thermometer of mass m and specific heat c is plunged at a temperature of T_i °C into a large container of liquid at T_0 °C. The overall heat transfer coefficient between the thermometer and water is K. The resulting temperature of the thermometer is given by

$$mc \frac{dT}{dt} = K(T_0 - T)$$

Express the *change in temperature* ΔT of the thermometer as a differential equation and solve to give this change directly as a function of time. What is the potential advantage of using changes in variables in dynamic equations as distinct from absolute values?

Answer $\Delta T = (T_0 - T_i)(1 - e^{-Kt/mc})$

3 (i) Use the Laplace transform method to solve the differential equations, with the given initial conditions, in the first problem, confirming the earlier solution in each case.

(ii) For the input/output relationship defined by the equation and boundary conditions

$$\frac{d^2x}{dt^2} + 5 \frac{dx}{dt} + 6x = u(t), \quad u(t) = 0 \text{ for } t < 0, \, 3 \text{ for } t \geq 0$$

$$x(0) = 2, \frac{dx}{dt}(t = 0) = 1$$

derive $x(t)$ using Laplace transforms and evaluate $x(1)$.

Answer $x(t) = 0.5 - 4\,e^{-3t} + 5.5\,e^{-2t}$, 1.045

4 A mechanical system is subjected to a steadily increasing force $u(t)$. The force increases at the rate of α N s^{-1} from an initial value of zero. Derive the general output position $x(t)$ if the equation defining the system dynamics is

$$\frac{d^2x}{dt^2} + 2.4 \frac{dx}{dt} + x = u(t)$$

Hence derive the response of the system to a unit step change in the force $u(t)$. Derive also from this the response of the system from rest to a sudden unit impulse in the applied force.

Answer $x(t) = \alpha(t - 2.4 + 2.618\,e^{-0.537t} - 0.218\,e^{-1.864t})$, $1 - 1.406\,e^{-0.537t} + 0.406\,e^{-1.864t}$, $0.755(e^{-0.537t} - e^{-2.864t})$

5 A simple feedback system is represented by the relationship

$$0.5 \frac{d^2r}{dt^2} + 6 \frac{dr}{dt} + r = 5 \frac{dc}{dt} + c$$

where $c(t)$ and $r(t)$ are the required and actual output values.
(i) What is the final error between input and output if the system is

subjected to (a) a unit step demand in c and (b) a unit ramp demand in c, i.e. $c = t$?

(ii) Express the output as an explicit function of time $r(t)$ in each case.

Answer

(i) 0, 1

(ii) $1 - 0.158 \ e^{-1.69t} - 0.842 \ e^{-11.831t}$, $0.935 \ e^{-0.169t} +$
  $0.071 \ e^{-11.831t} + t - 1$

6 The dynamic equation for two first order lags in series is

$$\frac{d^2x}{dt^2} + 4\frac{dx}{dt} + 3x = u(t)$$

where $u(t)$ is the input to the first lag and $x(t)$ is the final output.

(i) What is the response of this system to a unit impulse?

(ii) Using the answer to (i) and the convolution integral evaluate the value $x(t)$ at $t = 2$ s when the input has the form of a pulse of unit magnitude of duration of 1 s.

Answer

(i) $0.5(e^{-t} - e^{-3t})$

(ii) 0.109

7 Use the convolution integral to determine the output, as a function of time, of a first order system of unity steady state gain and unity time constant when the input $u(t) = A \sin 2t$ and evaluate this output when $t = 0.2$ s.

Answer $0.4A[e^{-t} + 0.5\sqrt{5} \ \sin \ (2t - \tan^{-1}2)]$, $0.037A$. (Angle in radians)

8 The voltage $v(t)$ across two particular points following operation of a switch in a voltage driven passive electrical circuit is given by the equation

$$\frac{d^2v}{dt^2} + 0.006\frac{dv}{dt} + 0.001v = 0.001e$$

where e is the input from a voltage source. Express this relationship as a transfer function. If $e(t)$ is at a constant value E determine $v(t)$.

Answer $v(t) = E[1 - 1.0045 \ e^{-0.003t} \ \sin(0.03145t + 1.476)]$

9 The displacement $x(t)$ of a mass under the action of an applied force $f(t)$ and restricted by a spring−damper combination is given by the dynamic equation

$$m\frac{d^2x}{dt^2} + \lambda\frac{dx}{dt} + kx = f$$

(i) If $m = 5$ kg, $\lambda = 5$ N s m^{-1} and $k = 100$ N m^{-1} how does the mass move if released from a position away from its equilibrium condition (i.e. released with the spring in an extended or compressed state)?

(ii) What is the transfer function between the input force and output movement?

(iii) A force $f(t)$ of constant magnitude 10 N is applied when the mass is initially at the equilibrium position. How far will it move in the next 0.2 s?

Answer

(i) Lightly damped oscillations,
(ii) $0.2/(s^2+s+20)$,
(iii) 0.035 m

10 A process involves using two tanks with the output from one flowing directly into the second. Each tank contains a constant amount of liquid and the flow rate through both tanks remains constant and equal. Both tanks are stirred and the change in a tank's output concentration c as a result of a change in input concentration w is given by

$$T\frac{dc}{dt} + c = w$$

where T is the tank time constant.

(i) What is the impulse response function relating output concentration to input changes for a single tank?

(ii) Using the convolution integral establish the impulse response for the unit of two tanks in series if they have respective time constants T_1 and T_2.

(iii) Using transfer functions determine the second tank output concentration change to impulse and step changes in the initial flow concentration into the first tank.

Answer

(i) $\dfrac{1}{T} (e^{-t/T})$

(ii) $\dfrac{1}{T_1-T_2} (e^{-t/T_1} - e^{-t/T_2})$

(iii) $\dfrac{1}{T_1-T_2} (e^{-t/T_1} - e^{-t/T_2})$, $\dfrac{1-T_1e^{-t/T_1}}{T_1-T_2} + \dfrac{T_2e^{-t/T_2}}{T_1-T_2}$

11 Fluid flows through a pipe from a reactor. The temperature as it leaves the reactor is Θ_r. The pipe is well lagged so that when the fluid passes a measurement probe in the pipe after 5 s its temperature is unchanged. However, because of the delay in reaching the point of measurement, the temperature measured Θ_m is actually that in the reactor 5 s earlier.

(i) What is the transfer function relating the measured temperature to the reactor temperature?

Following a change θ_i in the temperature of the principal reactant Θ_i there is a *change* in reactor temperature given by

$$T \frac{d\theta_r}{dt} + \theta_r = \theta_i$$

(ii) What is the transfer function $\theta_m(s)/\theta_i(s)$ giving the change in measured temperature?

(iii) What is $\theta_m(t)$ following a step change in θ_i of magnitude 10 °C if $T = 30$ s?

Answer

(i) e^{-5s}

(ii) $\dfrac{e^{-5s}}{1+Ts}$

(iii) $\mathcal{H}(t-5) \cdot (1 - e^{-(t-5)/T})$

2

Linear system modelling

Introduction

The ability to solve differential equations using techniques such as the Laplace transform is of no benefit in the study of dynamics and control unless a suitable and correct mathematical description of the system behaviour is made available. In fact, with the establishment of well defined methods of analysis and synthesis of mathematical expressions in the field of control, and their continuing extension, the major stumbling block is frequently that of obtaining the equations representing system behaviour in the first place. It may frequently be the case, with more complex or ill-defined situations, that only an approximate description of the system behaviour or one that is only valid over a limited operating range is possible. This tends to be the case with process systems which may involve chemical reactions tying together extensive mass and heat balance relationships. On the other hand it may be possible to 'model' simple mechanical or electrical systems with comparative ease. However, even in the latter case it is recognized that the modelling stage of a problem is not something which is usually seen as straightforward and it is always possible to come up against, or construe, a novel situation.

Similarities in the mathematical expressions representing the behaviour of quite different physical systems mean that we may frequently study one form of system by looking at the behaviour of its 'analogue'. Such similarity may also help in understanding the dynamics of a less familiar process by relating it to one which is better known to us.

Despite the difficulties which may arise in forming a mathematical model the reward is that, provided it can be constructed in a standard format, the methods of solution are applicable *to the model* and may subsequently be applied largely independently of the original system type. However, caution must be retained in recognition of the fact that the model may be only approximate or limited in range of applicability. The problems associated with poor modelling may be relieved by the use of feedback in control which has the effect of overcoming this uncertainty.

In all systems the basic underlying principle is that of 'conservation' or 'continuity'. This is true for both steady state and dynamic conditions and it gives rise to the laws of conservation of mass, momentum and energy. These may all be expressed by the single equation.

Accumulation within the system = total input − total output + generation within the system boundaries − consumption within the system boundaries

---

System inputs

Although not strictly a part of the system modelling process, which is independent of the system input, it is necessary to be able to express the input in standard forms to assess system behaviour under standard conditions. Indeed, non-random inputs can be expressed as summations of these standards. Consequently, because of the linearity property of the systems which we are considering, it is possible to evaluate the output resulting from defined inputs.

The standard non-periodic inputs are the impulse, the step and the ramp. They have the Laplace transforms shown in Chapter 1.

2.1 Laplace transforms of standard inputs

Use of the transfer function to derive a system's output requires also the Laplace transform of the particular input. Use the Laplace transform of the unit impulse and/or unit step to form the transform of the inputs in Fig. 2.1.

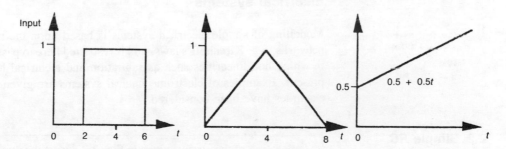

Figure 2.1 Examples of input forms

Solution Although one can always build up a transform through operations on the impulse in this case the use of the step as a basis for forming the Laplace transform of the various functions is adequate.

(i) The first graph of a *pulse* is comprised of two unit delayed steps

$$u(t) = \mathcal{H}(t-2) - \mathcal{H}(t-6)$$

so

$$u(s) = \frac{1}{s}(e^{-2s} - e^{-6s})$$

(ii) The *single saw tooth* input may be analysed in sections:

From $t = 0$ to 4 $u_1 = 0.25t$

From $t = 4$ to 8 $u_2 = 0.25t - 0.5(t-4)$

From $t = 8$ onwards $u_3 = 0.25t - 0.5(t-4) + 0.25(t-8)$

Thus the general expression is

$$u = 0.25t - \mathcal{H}(t-4)0.5(t-4) + \mathcal{H}(t-8)0.25(t-8)$$

The ramps may be treated as integrals of the corresponding delayed steps and

$$u(s) = \frac{1}{s}\left(\frac{0.25}{s} - \frac{0.5\,e^{-4s}}{s} + \frac{0.25\,e^{-8s}}{s}\right)$$

$$= \frac{1}{s^2}\left(0.25 - 0.5\,e^{-4s} + 0.25\,e^{-8s}\right)$$

(iii) The *second ramp* with non-zero value at $t = 0$ is simply a step of magnitude 0.5 superimposed with a ramp of $0.5t$. Thus, using the step as a basis again, the transformed expression for the input is

$$u(s) = \frac{0.5}{s} + \frac{1}{s}\left(\frac{0.5}{s}\right)$$

$$= \frac{0.5}{s^2}(1+s)$$

Electrical systems

Modelling of simple electrical systems is based upon the rules of electrical networks, e.g. Kirchhoff's laws, and is restricted here to passive *RCL* systems in which non-linearities such as saturation and electrical hysteresis are not present. Examples of electromechanical systems are given after mechanical examples have been considered.

2.2 Simple *RC* circuits

> For each of the circuits given in Fig. 2.2 derive the dynamic equations between the source (current or voltage) and the corresponding voltage or current associated with the capacitor.
>
> Express the relationships as transformed variables and a transfer function.

Solution (i) For the resistor and capacitor in *parallel* Kirchhoff's laws for current summation lead to

$$i = \frac{V}{R} + C\frac{dV}{dt}$$

Transforming using zero initial conditions, which will generally be assumed unless given otherwise,

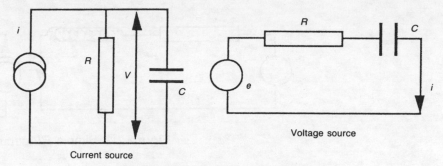

Voltage source

Current source

Figure 2.2 Simple *RC* circuits

$$i(s) = \frac{V(s)}{R} + CsV(s)$$

with the transfer function, for a current source input,

$$\frac{V(s)}{i(s)} = \frac{R}{1+RCs}$$

This is a first order system having the transfer function of a simple lag with a time constant RC and steady state gain R.

(ii) For the series combination the total voltage drop is V_R plus V_C. The current i is the same in each element being

$$C\frac{dV_C}{dt} = i$$

and

$$\frac{V_R}{R} = \frac{(e-V_C)}{R} = i$$

Differentiating this second equation and substituting from the first gives the relationship between the capacitor current and voltage source

$$RC\frac{di}{dt} + i = C\frac{de}{dt}$$

with input/output transfer function

$$\frac{i(s)}{e(s)} = \frac{Cs}{1+RCs}$$

The circuit time constant is RC.

2.3 Simple *RL* circuits

> Form the dynamic equations for the simple resistor–inductor combinations given in Fig. 2.3. Again express these as transformed variables and transfer functions.

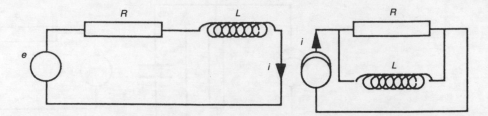

Figure 2.3 Simple *RL* circuits

Solution (i) Following similar reasoning, for the *series* combination

$$e = Ri + L \frac{di}{dt}$$

to give the transfer function

$$\frac{i(s)}{e(s)} = \frac{1/R}{1+Ls/R}$$

The time constant is L/R.

(ii) For the *parallel* combination the voltage V across each element is the same and we use summation of the currents at the nodes. Thus

$$i_R = \frac{V}{R} \text{ and } L \frac{di_L}{dt} = V \text{ and } i = i_L + i_R$$

Differentiating i and substituting for i_L and i_R leads to

$$\frac{1}{R} \frac{dV}{dt} + \frac{V}{L} = \frac{di}{dt}$$

with the system transfer function

$$\frac{V(s)}{i(s)} = \frac{sL}{1+Ls/R}$$

The time constant in this case is L/R.

2.4 Simple *RCL* circuit

Figure 2.4 shows a resistor–capacitor–inductor (*RCL*) circuit with a current source i. Derive the differential equation expressing the voltage across the capacitor in terms of the step input current i. If the initial conditions are 'zero', i.e. no applied voltage and no charge on the capacitor, derive the overall relationship in terms of Laplace transformed variables.

Solution This example shows one circuit using three passive elements. The order of the differential equation is correspondingly increased. Also initial

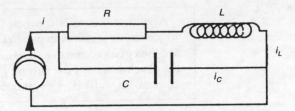

Figure 2.4 *RCL* circuit

conditions require careful consideration. The fundamental equations are

$$i = i_R + i_C = i_L + i_C \text{ as } i_R = i_L$$

and

$$V_C = V_R + V_L$$

Using the corresponding current−voltage relationships for the individual elements, as above, in these expressions leads to

$$V_C = i_L R + L \frac{di_L}{dt}$$

$$= R(i-i_C) + L \frac{d(i-i_C)}{dt}$$

$$= R(i-i_C) - L \frac{di_C}{dt} \text{ (as } i \text{ is a constant value for } t>0)$$

$$= Ri - RC \frac{dV_C}{dt} - LC \frac{d^2V_C}{dt^2}$$

to give the input/output form

$$LC \frac{d^2V_C}{dt^2} + RC \frac{dV_C}{dt} + V_C = iR$$

Now from the initial conditions $V_C = 0$. Because of the inductance there can be no initial finite flow of current i_L and at the instant the current source is applied this current flows to the capacitor so that $CdV_C/dt(0) = i(= I$, say). If this is taken into account when taking transforms of the dynamic equation then

$$LC \left[s^2 V(s) - \frac{I}{C} \right] + RCsV_C(s) + V_C(s) = i(s)R = \frac{IR}{s}$$

and

$$V_C(s) = \frac{RI}{s(LCs^2 + RCs + 1)} + \frac{I}{C(LCs^2 + RCs + 1)}$$

There is still a transfer function form but note that consideration of the system and its behaviour has led here to the initial condition on rate of change of the independent variable adding a further term. The system dynamic time constants are the same in both terms.

2.5 Simple *RCL* circuit

A voltage source is connected to the given *RCL* circuit in Fig. 2.5 which shows a switch in parallel with a capacitor. The switch is closed and the circuit comes to a steady condition. Derive the expression giving the development of the current *i* in the circuit when it is opened ($R = 20$ ohm, $L = 10$ H, $C = 0.005$ F, $e = 20$ V).

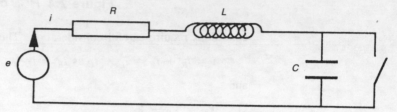

Figure 2.5 Switched *RCL* circuit

Solution Again the 'initial conditions' now appear explicitly in the solution. With the switch closed and a constant dc voltage source the voltage across the capacitor will be zero (short circuited). The current is steady and thus all the voltage drop is across the pure resistor. The initial conditions are

$$i(0) = \frac{e}{R}, \ V_C(0) = 0, \text{ and } V_L(0) = 0$$

After the switch is opened the general equations are

$$V_R + V_L + V_C = e$$
$$i = i_R = i_L = i_C$$

with $V_L = Ldi/dt$, $V_R = Ri$, $i = CdV_C/dt$.
Combining these equations gives

$$L\frac{di}{dt} + Ri + \frac{1}{C}\int i \, dt = e$$

Transforming and entering the non-zero initial condition gives

$$Lsi(s) - \frac{L}{R}e + Ri(s) + \frac{i(s)}{Cs} = e(s)$$

Rearranging, with $e(s) = e/s$, gives

$$i(s) = \frac{Cs(e/s + Le/R)}{LCs^2 + RCs + 1}$$

and, on entering the given values, this becomes

$$i(s) = \frac{2+s}{s^2+2s+20}$$

$$= \frac{(s+1) + 1}{(s+1)^2 + 19}$$

to give in turn on inversion

$$i(t) = e^{-t}\cos(\sqrt{19}.t) + \frac{1}{\sqrt{19}} e^{-t}\sin(\sqrt{19}.t)$$

$$= 1.026 \, e^{-t}\sin(\sqrt{19}.t + 1.345)$$

Note that the angle is expressed in radians.

Mechanical systems

Mechanical systems include a very wide range of everyday items which may frequently be reduced to a standard first or second order model. In combination with electrical components mechanical systems give rise in turn to electromechanical systems, e.g. motors and generators.

2.6 Mass–damper system

Figure 2.6 shows two alternative applications of force to a simple mass–damper system. In the first case force is applied directly to the mass which is separated from a fixed rigid surface by a light damper with damping coefficient λ. In the second case the restraining surface is absent and the force P is now applied to this end of the damper system. In this latter case the force is not now the independent variable but the velocity of movement of this part of the damper. If the system is initially at rest in each case, derive the relationship between the movement of the mass and the independent variable which is the forcing input.

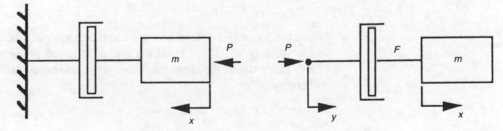

Figure 2.6 Forced mass–damper systems

Solution (i) For the *first configuration*, equating net applied forces to the acceleration force on the mass, or using d'Alembert's principle, gives the dynamic equation.

For applied external force $= P$ with damping force $= -\lambda \, dx/dt$ the equation of motion is

$$P - \lambda \frac{dx}{dt} = m \frac{d^2x}{dt^2}$$

i.e.

$$\frac{d^2x}{dt^2} + \frac{\lambda}{m}\frac{dx}{dt} = \frac{P}{m}$$

This may be expressed alternatively in terms of the velocity v

$$\frac{dv}{dt} + \frac{\lambda}{m}v = \frac{P}{m}$$

Continuing the use of Laplace transforms

$$\left(s^2 + \frac{\lambda}{m}s\right)x(s) = \frac{P(s)}{m}$$

The solution naturally depends on the function $P(t)$ but for a steady force, magnitude P, then $P(s) = P/s$ and

$$x(s) = \frac{P/m}{s^2(s+\lambda/m)}$$

to give

$$x(t) = P\left[\frac{t}{\lambda} - \frac{m}{\lambda^2}(1-e^{-\lambda t/m})\right]$$

(ii) The unconstrained elements in the *second figure* give rise to a different dynamic equation. The force transmitted across the damper is proportional to the relative velocity of the components so that balancing forces gives

$$P = \lambda(y' - x') = F$$

and

$$F - mx'' = 0$$

where the ' signifies differentiation with respect to time. The independent input is the velocity y' (say v_1) and the output required is the movement of the mass. There is thus the simple input/output relationship between the velocities (where $v_2 = x'$)

$$mv'_2 + \lambda v_2 = \lambda v_1$$

giving

$$\left(s + \frac{\lambda}{m}\right)v_2(s) = \frac{\lambda}{m}v_1(s)$$

If the input is a steady velocity V (achieved instantaneously assuming the components of the damper are of zero mass), then $v_1(s) = V/s$ and

$$v_2(s) = \frac{\lambda}{m}\frac{V}{s(s+\lambda/m)}$$

$$= \frac{V}{s(1+sT)} \text{ where } T = \frac{m}{\lambda}, \text{ the time constant}$$

Then

$$v_2(t) = V(1 - e^{-t/T})$$

with displacement

$$x(t) = V(t - T + Te^{-t/T})$$

2.7 Spring–damper system

A light spring–damper coupling may be composed of parallel or serially connected components. Although one end of the system may be fixed or connected to an inertial mass, the modelling and movement may be initially investigated when there is movement at each end of the system and no mass in the system. (One of the velocity terms may be set to zero by fixing that end.) Derive a relationship in each case in Fig. 2.7 between the difference in velocity between the endpoints of the system and forces applied to the system.

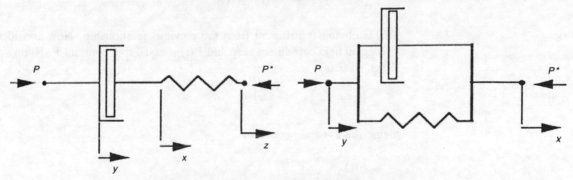

Figure 2.7 Forced spring–damper systems

Solution With these free systems, relationships may be expressed in terms of the relative velocities of the physical extremes, say $y' - z'$ in the series arrangement and $y' - x'$ in the second parallel arrangement. Denote the derivatives y', x', z' by v_1, v_2, v_3 respectively.

(i) Considering forces through the *series* connection of the spring (stiffness k) and damper (viscous friction coefficient λ),

$$P - \lambda(v_1 - v_2) = 0$$
$$\lambda(v_1 - v_2) - k(x - z) = 0$$
$$k(x - z) - P^* = 0$$

Thus $P = P^*$ and eliminating the intermediate variable v_2, which is not required in the solution, using differentiation of the third equation leads to

$$P' + \frac{k}{\lambda} P = k(v_1 - v_3)$$

If one end is held so that $v_3 = 0$ and the other end is moved in with constant velocity so that $v_1 = V$, a constant, then the force P will rise, assuming no initial compression and that linearity is maintained, according to

$$\left(s + \frac{k}{\lambda}\right) P(s) = k \, \frac{V}{s}$$

i.e.

$$P(t) = \lambda V (1 - e^{-tk/\lambda})$$

tending to a value of λV.

(ii) For the *parallel* arrangement there is no intermediate velocity and considering the condition at each end of the spring−damper combination gives

$$P - k(y-x) - \lambda(v_1 - v_2) = 0$$
$$k(y-x) + \lambda(v_1 - v_2) - P^* = 0$$

Once again $P = P^*$ and the force−velocity relationship is now, on removing the explicit position variables by differentiation,

$$(v_1' - v_2') + \frac{k}{\lambda} (v_1 - v_2) = \frac{P'}{\lambda}$$

This is obviously different from the previous relationship. Now consider if v_2 is again held constant at zero and $P(t)$ is equal to a constant P. Noting that $P'(s) = sP(s)$,

$$\left(s + \frac{k}{\lambda}\right) v_1 = \frac{sP}{\lambda s}$$

giving the dynamic equation

$$v_1(t) = \frac{P}{k} \, e^{-kt/\lambda}$$

and

$$y = \frac{P}{k} (1 - e^{-kt/\lambda})$$

tending to P/k when the spring force balances the force P and the system is at rest.

2.8 Mass−spring−damper combination

Many mechanical systems can be represented as a combination of one or more mass, spring and damper configurations. Such a representation may be used for suspension systems, machine tool vibrations, linkage of vehicles, etc. One such arrangement is shown in Fig. 2.8. Relate the movement of the mass to the applied force P, forming an overall transfer function between this and the mass displacement. How does this system respond to a step force of 20 N? (Use values of $m = 200$ kg, $\lambda = 100$ N s m^{-1}, $k = 600$ N m^{-1}.)

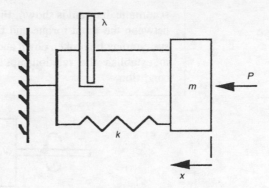

Figure 2.8 Mass−damper system

Solution This mechanical system gives rise to the standard second order system dynamic equation. Summing the applied force, the damping force and the spring force, all applied directly to the mass, with the d'Alembert term (for mass acceleration), respectively, gives

$$P - \lambda x' - kx - mx'' = 0$$

i.e.

$$mx'' + \lambda x' + kx = P$$

Hence the required transfer function is

$$\frac{x(s)}{P(s)} = \frac{1}{m\,s^2 + \lambda s + k}$$

For a constant force of 20 N and with the given system parameters,

$$x(s) = \frac{20}{s(200s^2 + 100s + 600)}$$

$$= \frac{1}{30}\,\frac{3}{s(s^2 + 0.5s + 3)}$$

In this form inversion is easy by direct use of the table of transforms,

$$x(t) = \frac{1}{30}\,[1 - 1.01\,e^{-0.25t}\sin(1.714t + 1.426)]$$

where the sine function is here expressed in radians. The final steady state displacement is $x(t) = 1/30$ m, i.e. P/k.

2.9 Rotary inertia system

Torque is applied through a stiff shaft to a viscous coupling having a drive side moment of inertia of J_1 N m s^2. The driven shaft has an effective moment of inertia J_2 N m s^2 and the transmitted torque is given by $\lambda(\omega_1 - \omega_2)$ where ω is the corresponding shaft speed (rad s^{-1}) and λ is the coefficient of viscous friction (N m s^{-1}). The physical schematic diagram is shown, Fig. 2.9. Establish the overall relationship between the input torque and the output speed, showing also how the system may be broken down and the Laplace transformed variables used to establish the relationships for deviation from a steady running condition.

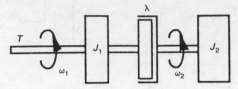

Figure 2.9 Rotational inertia system

Solution The system may be conveniently broken down into two parts by pulling apart at the two elements of the viscous coupling. Taking the *input side* and designating the transmitted torque at the coupling as T_{12} leads to the torque equation

$$T = J_1\omega_1' + T_{12}$$

For the *output side*

$$T_{12} = J_2\omega_2'$$

The transmission of torque is given by

$$T_{12} = \lambda(\omega_1 - \omega_2)$$

Combining these to eliminate T_{12} and ω_1 leads to the input/output relationship

$$\frac{J_1J_2}{\lambda}\omega_2'' + (J_1 + J_2)\omega_2' = T$$

Transformed variables may be used during the derivation of the overall dynamics. Modelling by way of deviations from a steady speed and torque and subtracting the steady condition equations (when the time derivatives of the speeds will be zero) gives

$$\delta T = J_1\delta\omega_1' + \delta T_{12}$$
$$\delta T_{12} = J_2\delta\omega_2'$$
$$\delta T_{12} = \lambda(\delta\omega_1 - \delta\omega_2)$$

with now all-zero initial conditions in the new variables. This makes it possible to form an overall transfer function in terms of input and output changes. Eliminating intermediate variables as before (a step which could be done after taking transforms),

$$\frac{J_1J_2}{\lambda}\delta\omega_2'' + (J_1 + J_2)\delta\omega_2' = \delta T$$

and the transfer function is

$$\frac{\delta\omega_2(s)}{\delta T(s)} = \frac{\lambda}{s[J_1J_2s + \lambda(J_1 + J_2)]}$$

The final time response then depends on the forcing disturbance $\delta T(t)$.

2.10 Moment of inertia with spring and damping

A swing door of moment of inertia about its hinges J is fitted with a spring—damper system to ensure that it returns to its central closed position. If the angle of the door into a room is θ the spring supplies a closing torque of $k\theta$ and the damper limits the speed of closure by supplying a torque $-\lambda\theta'$, i.e. proportional to the velocity of the door but in the opposite sense to its angular velocity.

(i) Derive the transfer function between an applied torque T and the angle of opening θ.

(ii) If the coefficient of the damper is to be adjusted so that when released from rest in an open position the door just closes without overshoot in minimum time, express the free closure of the door $\theta(t)$ as a function of J, k and time t.

Solution (i) The torques acting on the door tending to force it *to the closed position* $\theta = 0$ are

fom the spring	$k\theta$
from the damper	$\lambda\theta'$
from the applied torque	$-T$
d'Alembert torque	$J\theta''$

The applied torque has to overcome the other forces in opening the door. From the conservation equation

$$J\theta'' + \lambda\theta' + k\theta - T = 0$$

giving the transfer function

$$\frac{\theta(s)}{T(s)} = \frac{1}{J}\frac{1}{(s^2 + s\lambda/J + k/J)}$$

(ii) To determine the required condition of critical damping the damping coefficient c is unity. Compare the above expression with that for the 'standard second order system',

$$\frac{\omega_n^2}{s^2 + 2c\omega_n s + \omega_n^2}$$

Then $\omega_n = \sqrt{(k/J)}$ and $2c\omega_n = \lambda/J$. For critical damping $c = 1$. Eliminating ω_n gives the relationship $\lambda = 2\sqrt{(kJ)}$. Because the closure from an initial opening Θ is required with $T = 0$ and $\Theta' = 0$ the solution of

$$J\theta'' + \lambda\theta' + k\theta = 0$$

is required with these initial conditions. Solving by Laplace transforms leads to

$$J[s^2\theta(s) - s\Theta] + \lambda[s\theta(s) - \Theta] + k\theta(s) = 0$$

i.e. with the substitution for λ being made

$$\theta(s) = \Theta\left\{\frac{s}{[s + \sqrt{(k/J)}]^2} + \frac{2\sqrt{(k/J)}}{[s + \sqrt{(k/J)}]^2}\right\}$$

This gives on inversion

$$\theta(t) = \Theta[1 + \sqrt{(k/J)t}]e^{-\sqrt{(k/J)t}}$$

with the general normalized form $\theta(t)/\Theta$ *vs* $\sqrt{(k/J)t}$ of Fig. 2.10.

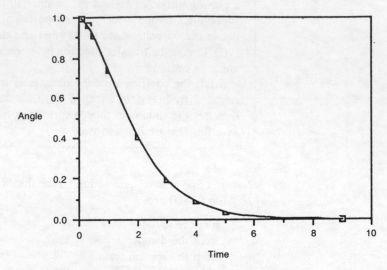

Figure 2.10 Closing of damped door

Note the double root at critical damping and the absence of overshoot in the response.

2.11 Mechanical and electrical analogues

Electric analogues are frequently used to study other systems. Figure 2.11 shows an electric circuit to simulate the behaviour of the door in the previous question. The voltage V across the capacitor represents the angular velocity. If the current source represents the applied torque establish the equivalent dynamic equation and transfer function for this circuit. If the damping coefficient is the same in both cases what is the relationship between the parameters of the two physical systems?

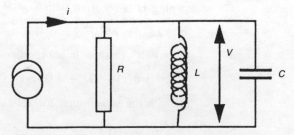

Figure 2.11 Electrical analogue for door behaviour

Solution For the electric circuit with the elements in parallel the sum of the component currents will be equal to the current source input. Taking the resistor, inductor and capacitor respectively and summing the currents gives

$$i_R + i_L + i_C = i$$

i.e.

$$\frac{V}{R} + \frac{1}{L} \int V dt + C \frac{dV}{dt} = i$$

Taking Laplace transforms

$$\left(\frac{1}{R} + \frac{1}{Ls} + Cs \right) V(s) = i(s)$$

Now, if $V \equiv d\theta/dt$ then $V(s)/s \equiv \theta(s)$ and we may write this equation as

$$\left(s^2 + \frac{s}{RC} + \frac{1}{LC} \right) \frac{V(s)}{s} = \frac{i(s)}{C}$$

Comparing this with the above example the corresponding natural frequencies are $\sqrt{(k/J)}$ and $\sqrt{(1/LC)}$. The damping coefficients in the two cases are $c = \lambda/[2\sqrt{(kJ)}]$ and $\sqrt{(LC)}/2RC$. For these values of c to be equal then

$$\frac{1}{R} \sqrt{\left(\frac{L}{C} \right)} = \frac{\lambda}{\sqrt{(kJ)}}$$

Electromechanical systems

As mentioned above, electrical components and mechanical systems are combined to give electromechanical systems, e.g. motors and generators. As such growth in a system occurs the order of the system, that of the differential equation representing it, also increases. However, if the time constants of the combined system are widely different in magnitude then subsequent order reduction may be achieved by simplification based on removing the shorter time constants, the behaviour being dominated by the longer time constants (slower modes) in the system. Alternatively, if the fast initial reaction of a system is required, these modes must be retained.

2.12 Transfer function for an electromechanical system

A fundamental electromechanical system is the servomechanism used, for example, to position a link of a robot arm, a radar tracking dish, a control surface in aircraft, process control valves, and machine tools. The principles of control in these devices are considered in later examples but the combination of a simple electrical machine driving an inertial load is illustrated here. The overall dynamics are built up by consideration of the elements of the system.

Figure 2.12 is a representation of a simple servomechanism. Establish the overall transfer function between a demanded change in the output shaft position θ_d and the actual movement made by that shaft θ_o. Assume that the electrical side of the motor including the amplifier has an overall time constant of T_a and gain K_a in producing the motor torque; the load at the motor output has moment of inertia J and damping coefficient λ. The voltage to the amplifier is proportional to the difference between the demanded and output shaft positions such that $V = K(\theta_d - \theta_o)$.

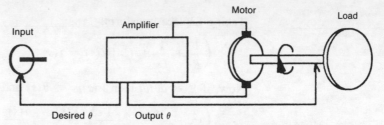

Figure 2.12 Servomechanism with position feedback

Solution Consider the following division of the system. The input voltage to the amplifier is

$$V = K(\theta_d - \theta_o)$$

giving the transform equation

$$V(s) = K[\theta_d(s) - \theta_o(s)]$$

The torque Q produced by the combination of amplifier and motor is by a gain K_a and with a lag given by the time constant T_a. Thus the torque–voltage transfer function is

$$\frac{Q(s)}{V(s)} = \frac{K_a}{1+T_a s}$$

The torque to output shaft position dynamic equation is

$$Q = J\frac{d^2\theta_o}{dt^2} + \lambda\frac{d\theta_o}{dt}$$

giving the transfer function

$$\frac{\theta_o(s)}{Q(s)} = \frac{1}{Js^2 + \lambda s}$$

Eliminating the intermediate variables from these transformed equations leads to

$$\frac{\theta_o(s)}{\theta_d(s)} = \frac{KK_a}{(1+T_a s)(Js^2+\lambda s) + KK_a}$$

$$= \frac{KK_a}{T_a Js^3 + (T_a\lambda+J)s^2 + \lambda s + KK_a}$$

Hydraulic systems

Hydraulic systems have a complexity of model which depends on the degree to which properties such as fluid compressibility are taken into account. However, if these terms are omitted then these systems give rise to model equations of standard form. This is also true of pneumatic systems, although in this case the properties of the fluid will have a more marked effect on both the system behaviour and on the results predicted by an oversimplified model. Such systems may also be subject to minor leakage flows, e.g. within the actuators, which lead to additional model terms which it is often hard to quantify.

2.13 Hydraulic ram action

The working of a hydraulically powered baling press used to compress a loose material into denser blocks is represented by the simplified mass—spring—damper system shown in Fig. 2.13. The supply pressure p_s and parameters at the given stage of baling are assumed to be constant. The flow F into and out of the main ram chambers is governed by the spool valve position x and the load pressure p_l, the difference in pressures across the ram ($p_l = p_1 - p_2$), according to the equation,

$$F = C\left[(x_0+x)\sqrt{\left(\frac{p_s-p_l}{2}\right)} - (x_0-x)\sqrt{\left(\frac{p_s+p_l}{2}\right)}\right]$$

where x is the movement from the central 'no-flow' position and x_0 is a constant (the 'underlap') of the valve. For small movements x of this valve derive the corresponding movement of the ram y expressed as a transfer function, explaining how the ram movement for a short duration opening of this valve would be obtained.

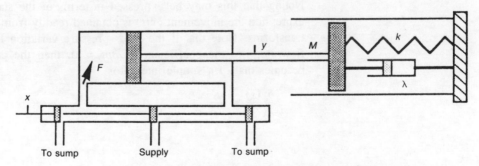

Figure 2.13 Hydraulic press schematic

Solution The terms which vary during a move are the position x and the load pressure p_l. Consider small changes from a stationary equilibrium position. Then linearization of the governing equation given in the question is achieved through

$$\delta F = \frac{\partial F}{\partial x}\,\delta x + \frac{\partial F}{\partial p_l}\,\delta p_l$$

where the partial differentials are evaluated at the initial condition F_0 (equal to zero if there is no leakage past the piston) and p_{l0} are constants in the linear representation. If A is the cross-sectional area of the main ram then the speed of the mass at the output shaft is related to the instantaneous flow by

$$\delta F = A\,\frac{d(\delta y)}{dt}$$

The force applied across the ram balances the forces, including d'Alembert's force, in the mass−spring−damper system so that

$$A\,\delta p_l = M\,\frac{d^2(\delta y)}{dt^2} + \lambda\,\frac{d(\delta y)}{dt} + k(\delta y)$$

Eliminating by algebra the intermediate variable δp_l from the equations and representing the differential terms (for convenience of expression) by

$$\frac{\partial F}{\partial x} = AK \quad \text{and} \quad \frac{\partial F}{\partial p_l} = -\frac{A^2 K}{b}$$

leads to

$$\frac{d(\delta y)}{dt} + \frac{K}{b}\left[M\,\frac{d^2(\delta y)}{dt^2} + \lambda\,\frac{d(\delta y)}{dt} + k(\delta y)\right] = K\delta x$$

As deviations from an initial equilibrium condition are being used the initial conditions on the variables are zero and the transfer function below results directly on taking Laplace transforms,

$$\frac{\delta y(s)}{\delta x(s)} = \frac{b/M}{s^2 + (b + \lambda K)/MK.s + k/M}$$

Noting that this may be expressed in terms of the standard second order expression the movement $\delta y(t)$ is obtained readily from the look-up table of transforms. Note that if the load−pressure variation has no effect on the flowrate, i.e. the coefficient $\partial F/\partial p_l = 0$, then the effect of this system becomes that of a simple integrator

$$\frac{\delta y(s)}{\delta x(s)} = \frac{K}{s}$$

Process systems

Under the heading of process systems we consider those which are not described in direct terms of the electrical, mechanical and hydraulic types above. Thus they include reference to heat transfer, mixing, possibly reactions, and frequently variables which are 'distributed' in nature. The variables are then dependent on more than one independent variable, most commonly a position as well as time. In certain cases they lead to linear time-invariant differential equations, but now they are partial differential equations. However, it is still possible to effect a solution by the use of Laplace transforms and transfer functions as well as through other standard methods of solution. These models may be simplified by making gross assumptions, such as treating flow through a heated pipe as flow through a jacketed stirred vessel. In many cases the accuracy of representation is quite adequate for use in a control design study.

2.14 Process system and analogue

A mixer is driven by means of a fluid coupling, with viscous torque coefficient λ, from a constant speed motor via stiff shafts as in Fig. 2.14. The effect of the turning of the blades in the mixer contents is to produce a viscous drag torque on the connecting shaft and the equivalent moment of inertia on the coupling output shaft is J. Assuming that the motor speed remains constant, derive an electrical analogue which could be used to study the effect of sudden loads being placed on the system as a result of additions of extra material to the mixer placing in turn additional torque loading on its blades. Assume also that during these additions the moment of inertia and viscous drag coefficients are unchanged.

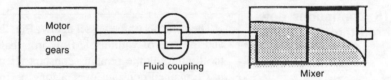

Figure 2.14 Mixer with fluid drive

Solution   To establish the analogue it is necessary first to understand and derive the dynamic equation for the original system. The transmitted torque through the drive is balanced by the drag in the mixer, acceleration torque in the mixer and the disturbance torque T_L caused by material addition. If the constant motor speed is ω_d, the mixer speed is ω_m and the effective coefficient of viscous drag for the mixer is b, then

$$\text{Transmitted torque} = \lambda(\omega_d - \omega_m)$$

$$= b\omega_m + J\frac{d\omega_m}{dt} + T_L$$

i.e.

$$J\frac{d\omega_m}{dt} + (b+\lambda)\omega_m = -T_L + \lambda\omega_d$$

For deviations from steady running, at which $T_L = 0$, $d\omega_m/dt = 0$ and the torque transmitted is $\lambda(\omega_d - \omega_m)$, there is a speed change $\delta\omega_m$ caused by the extra load T_L. (The extra load causes a reduction in speed.) Subtracting the steady state equation from the general equation gives the dynamic equation in terms of these deviations,

$$J \frac{d(\delta\omega_m)}{dt} + (b+\lambda)\delta\omega_m = -T_L$$

The term $-T_L$ is the *input* and the change $\delta\omega_m$ is the *output*. The straight electrical analogue equivalents are current for the torque and voltage for the velocity. From earlier considerations of passive electrical systems the analogue circuit is shown in Fig. 2.15 with equivalent dynamic equation

$$C \frac{dV}{dt} + \frac{V}{R} = i$$

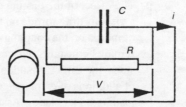

Figure 2.15 Electrical analogue for mixer system

2.15 Continuous flow stirred tank

The contents of the vessel below, Fig. 2.16, are well mixed at all times and the rate of volumetric flow in is balanced by the flow out so that the liquid volume remains constant. If the inlet flow concentration ρ_i has remained constant over a long period of time and is then subject to a step change in concentration $\delta\rho_i$ of one component, how does the outlet concentration ρ vary with time?

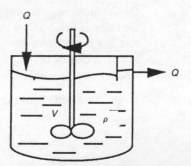

Figure 2.16 Continuous flow stirred tank

Solution This is about the simplest form of a continuous flow stirred tank system. The dynamic equation may be written directly in terms of the deviations from the initial steady conditions. Redefine the variable ρ as the *change* in concentration in the vessel following the introduction of the step change in inlet concentration $\delta\rho_i$. Directly applying the conservation equation, i.e. inflow = accumulation + outflow,

$$Q\delta\rho_i = V\frac{d\rho}{dt} + Q\rho$$

i.e.

$$\frac{d\rho}{dt} + \frac{Q}{V}\rho = \frac{Q}{V}\rho_i$$

For a step input ρ_i from zero initial conditions (in terms of the deviations) the transformed equation is

$$\rho(s) = \frac{Q\rho_i}{V}\left[\frac{1}{s(s+Q/V)}\right]$$

$$= \frac{\rho_i}{s(1+V/Q.s)}$$

This is a first order relationship with time constant V/Q. On inversion this yields the *change* in concentration,

$$\rho(t) = \rho_i(1 - e^{-Q/V.t})$$

and eventually the tank and output concentration becomes equal to the input concentration.

2.16 Fluid mixers in series

Figure 2.17 shows two perfectly mixed vessels used in the preparation of a chemical solution. A pure solvent flows at a constant rate Q of 0.01 m^3 s^{-1} into the top vessel which always contains a constant volume, 2 m^3, of solvent. It flows from here into the lower tank. A soluble solid is added to the upper tank where it dissolves rapidly without any increase in the solution volume. The solid is added at the rate shown in the figure, over a time of 60 s, a total of 1 kg being added. Initially the top vessel is full of fresh solvent and the bottom vessel is empty.

Establish a differential equation relating the concentration of solids in the solution in the first vessel to the rate of addition of solids. Via the formation of transfer functions determine the mean concentrations of the solutions in the two vessels after a total elapsed time of 140 s.

Solution Looking at the addition of solids into the top tank it is seen that the *total* reaches 1 kg after 1 minute. It will actually be more convenient to work in time units of minutes rather than SI units of seconds with the numerical values in this question. The fluid flow rate is then 0.6 m^3 per minute. The total added stays at this value so there is no further addition after this time.

The *input* function is thus a constant 1 kg per minute over the first minute only and this may be expressed as

$$u(t) = 1 - \mathcal{H}(t-1)$$

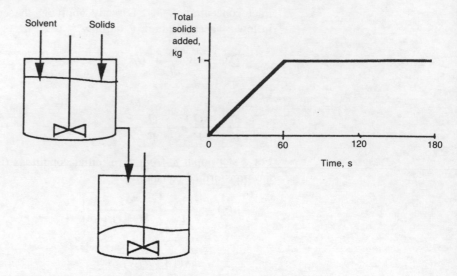

Figure 2.17 Mixed tanks in series

The mass balance on the solids content in the *first* vessel gives

$$2\,\frac{dc_1}{dt} = u - 0.6c_1$$

i.e.

$$\frac{dc_1}{dt} + 0.3c_1 = 0.5[1 - \mathcal{H}\,(t-1)]$$

The transformed equation, with zero initial conditions on c_1, is

$$c_1(s) = \frac{0.5}{s(s+0.3)}\,(1 - e^{-s})$$

$$= \frac{1.667}{s(1+3.333s)}\,(1 - e^{-s})$$

and the solution is

$$c_1(t) = 1.667[(1 - e^{-0.3t}) - \mathcal{H}\,(t-1)(1 - e^{-0.3(t-1)})]$$

After 140 s, 2.333 minutes,

$$c_1 = 0.290 \text{ kg m}^{-3}$$

For the *second*, lower, vessel which just collects the liquid from the first vessel after starting empty, the volume V collected is simply flowrate × time, i.e. $0.6t$. The total amount of solids C_2 is

$$\int 0.6c_1(t)dt \quad \text{or} \quad C_2(s) = \frac{0.6}{s}\,c_1(s)$$

from the expression for $c_1(s)$ or $c_1(t)$

$$C_2 = 1.667 \left\{ t + \frac{e^{-0.3t}}{0.3} - \frac{1}{0.3} - \mathcal{H}(t-1) \right.$$

$$\left. \left[(t-1) + \frac{e^{-0.3(t-1)}}{0.3} - \frac{1}{0.3} \right] \right\}$$

After 140 s, 2.333 minutes, $V = 1.4 \text{ m}^3$ and $C_2 = 0.700$ kg. The concentration c_2 at this time is 0.700/1.400, i.e.

$$c_2 = \textbf{0.200 kg m}^{-3}$$

2.17 Heat flow and electrical analogue

A strengthened insulating panel is made of three parallel plates of high strength but also of high heat conductivity separated by layers of material having conductive heat transfer coefficients of 0.1 kW K^{-1} m^{-2} but low heat capacity. The plates themselves have heat capacity of 20 kJ K^{-1} m^{-2} of contact surface. The surrounding atmosphere at the outer plate is at 0 °C and the inner plate is exposed to a reaction mixture which rises rapidly to a temperature of 150 °C and remains there. Prior to this all layers are at 0 °C. Derive a differential equation relating the temperature of the middle plate to the reaction mixture temperature.

Show also an electrical circuit giving an analogue of this thermal system.

Solution The situation for this example is shown in Fig. 2.18. The outer plates remain at constant temperatures because of their high conductivity and the presence of the constant temperature heat source and sink at 150 °C and 0 °C respectively. Because of their stated low thermal capacity the layers of insulation are considered not to accumulate heat.

The heat flow equations are, per m^2 of interface,

Insulation layer 1 $\qquad Q_1 = 0.1(150 - T_1)$

Central plate $\qquad Q_1 = 20 \dfrac{dT_1}{dt} + Q_2$

Insulation layer 2 $\qquad Q_2 = 0.1(T_1 - 0)$

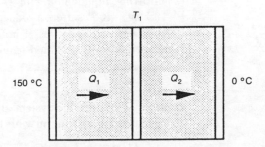

Figure 2.18 Heat flow system

Combining these to give the dynamic equation in T_1 gives

$$20 \frac{dT_1}{dt} + 0.1T_1 = 15 - 0.1T_1$$

i.e.

$$20 \frac{dT_1}{dt} + 0.2T_1 = 15$$

The actual rise in temperature in the central plate will depend on its initial value. Assuming an initial value of 0 °C

$$T_1 = 75(1 - e^{-0.01t})$$

The equivalent circuit (showing analogous variables) is as shown in Fig. 2.19. The variable current (heat flow) is divided between the resistor (second layer of insulation) and the capacitor (central sheet heat capacity).

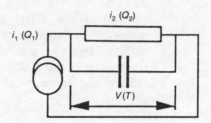

Figure 2.19 Electrical analogue for heat flow system

2.18 Continuous stirred tank reactor

Figure 2.20 is of an ideal 'continuous stirred tank reactor'. It is perfectly mixed and thus has a uniform temperature T throughout its contents at a given time. The heat of reaction, i.e. generated within the reactor, is small and the jacket is really intended to cope with its removal. The heat balance model may consider this heat of reaction element to be negligible. The coolant in the jacket flows at a sufficiently fast rate so that the jacket maintains a constant temperature T_j of 20 °C and the heat transfer coefficient K between the vessel contents and the jacket is 200 W m^{-2} K^{-1}. Effective heat transfer area A is 1.5 m^2. The inflowing liquid at a rate Q of 2 kg s^{-1} and heat capacity C_p of 4 kJ kg^{-1} K^{-1} suffers a sudden change in temperature from T_1 (80 °C) to T_2 (90 °C). The vessel holds a mass M of 100 kg. How does the temperature of the liquid leaving the vessel change with time?

Solution Consider the system at its initial steady state, vessel temperature T_0. Balancing heat flow in with heat flow out leads to

$$C_p Q T_1 = C_p Q T_0 + (T_0 - T_j)AK \dots\dots\dots\dots\dots\dots\dots [i]$$

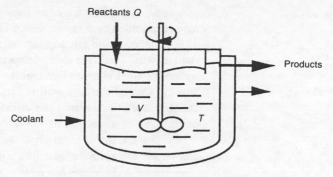

Figure 2.20 Continuous stirred tank reactor

After the inlet temperature change the general equation for transient temperature T becomes

$$C_p Q T_2 = C_p Q T + (T - T_j)AK + MC_p \frac{dT}{dt} \quad \dots\dots\dots\dots\dots \text{[ii]}$$

Subtracting [i] from [ii] gives

$$C_p Q(T_2 - T_1) = C_p Q(T - T_0) + (T - T_0)AK + MC_p \frac{d(T - T_0)}{dt}$$

Note that as $dT_0/dt = 0$ the last term above is also $MC_p \, dT/dt$. Writing the temperature difference $T - T_0$ as θ and substituting in the given values yields

$$\frac{400}{8.3} \frac{d\theta}{dt} + \theta = \frac{80}{8.3}$$

Solution, by transforms or otherwise, results in

$$\theta = 9.64(1 - e^{-0.0208t})$$

Solving equation [i] gives $T_0 = 77.83 \,°C$ so that

$$T = 87.47 - 9.64 \, e^{-0.0208t} \,°C$$

Because of the extra heat removed by the jacket on account of the higher temperature the vessel temperature does not increase by the full 10 °C. At the same time the cooling capacity is insufficient to reduce significantly the effect of the temperature rise.

2.19 Distributed system model

This example is intended to illustrate the ease with which complexity arises in the case of what appears to be a quite simple 'process' system. It also illustrates, however, the further use of the continuity equation, the use of quite severe simplifications, the use of the Laplace transform with partial differential equations and the formation of a transfer function for this 'distributed' system. Frequently, only by the use of simplifying assumptions can solutions be obtained in this way.

A simple heat exchanger as shown in Fig. 2.21 consists of a single tube surrounded by a steam heated jacket. Derive the governing partial differential equation relating the outlet temperature T_0 to a uniform step change in the shell side steam temperature T_j. Assume that the dynamics within the shell side are fast and the temperature remains uniform, the heat capacity of the tube wall is negligible, and the temperature within the tube is uniform across the direction of flow at each position along the tube in the direction of flow.

Derive the transfer function between these two variables (T_0 and T_j) and obtain the solution as a function of time.

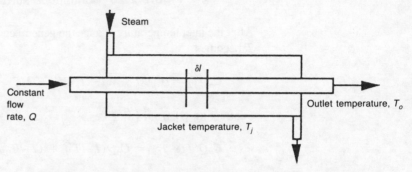

Figure 2.21 Distributed parameter heat flow system

Solution Define the properties of the fluid in the tube which are considered to remain constant as density, ρ, specific heat, c, heat transfer coefficient between wall and fluid, h, with internal tube diameter D, volumetric flow Q, and temperature at position l along the tube the variable $T(l, t)$. Taking a small time interval δt and applying the conservation equation to the element of length δl to equate the heat accumulation within the element to the net heat flow gives the series of terms

$$\delta l \, \frac{\pi D^2}{4} \, \rho c \, \frac{\partial T}{\partial t} \, \delta t = Q\rho c T \delta t - Q\rho c \left(T + \frac{\partial T}{\partial l} \, \delta l\right) \delta t + hD\pi\delta l (T_j - T)\delta t$$

$$= -Q\rho c \, \frac{\partial T}{\partial l} \, \delta l . \delta t + h\pi D \delta l (T_j - T)\delta t$$

This simplifies to

$$\frac{\pi D^2}{4} \, \rho c \, \frac{\partial T}{\partial t} = -Q\rho c \, \frac{\partial T}{\partial l} + h\pi D(T_j - T)$$

Combining the constant parameters in the groups $\tau = D\rho c / 4h$ and $\alpha = Q\rho c / h\pi D$ yields the equation for $T(l, t)$ as

$$\tau \, \frac{\partial T}{\partial t} + \alpha \, \frac{\partial T}{\partial l} = (T_j - T)$$

Now it is clear that T is a function of two independent variables and this is a characteristic of distributed parameter systems. It is only possible to proceed using the methods of the above examples if we make *substantial assumptions*. Assuming initial conditions of a constant temperature throughout the length of the tube, and the response of the fluid temperature to a *change* in jacket temperature of T_j, the *change* in fluid temperature θ is expressed by

$$\tau \frac{\partial \theta}{\partial t} + \alpha \frac{\partial \theta}{\partial l} = (T_j - \theta)$$

Using Laplace transforms with the time variable

$$(\tau s + 1)\theta(s) + \alpha \frac{d\theta(s)}{dl} = T_j(s)$$

gives $\theta(s)$ as a function of position l,

$$\theta(s) = \frac{T_j(s)}{(1+\tau s)} (1 - e^{-(\tau s + 1)l/\alpha})$$

With interest restricted to the output temperature at $l = L$, i.e. the change θ_0 in temperature T_0, the transfer function is

$$\frac{\theta_0(L, s)}{T_j(s)} = \frac{1}{(1+\tau s)} (1 - e^{-(\tau s + 1)L/\alpha})$$

For a step change of magnitude T_j

$$\theta_0(L, t) = T_j[(1 - e^{-t/\tau}) - e^{-L/\alpha} \, \mathcal{H}(t - t_0)(1 - e^{-(t-t_0)/\tau})]$$

where $t_0 = \tau L/\alpha$.

Note the limitations in using this method although it may give advantage in obtaining comparatively rapid approximate models and solutions which can be incorporated in a linear system analysis more easily than the full solution of the partial differential equations. Numerical methods, rather than analytical solutions, are likely to be required for more general distributed parameter systems.

Problems

1 The emf to a circuit is built up in three discrete steps of 5 V at 0.1 s intervals and is then switched off after a further 0.1 s. Sketch this emf as a function of time and by considering it as a summation of individual inputs deduce the Laplace transform of this input.

Answer $\frac{5}{s}(1 + e^{-0.1s} + e^{-0.2s} - 3e^{-0.3s})$

2 A true step input is in many instances not possible and the value may ramp up over a short period of time to its final value and then remain constant. At times which are considerably greater than the period taken for an input to reach its steady value the effect on the system may be adequately taken to be that of a true step. A 'step input' force F is modelled as a ramp of 10 N s^{-1} for 0.5 s followed by a constant value after this time of 5 N.

(i) By considering how this force may be made up of separate components determine its Laplace transform.

(ii) The force is applied to a mass–damper system initially at rest. The velocity $v(t)$ of this is governed by the equation

$$2\frac{dv}{dt} + v = F$$

What is the Laplace transform of the velocity of the mass?

(iii) Calculate the velocity of the mass at time $t = 2$ s based on (a) the input as described and (b) a true step at $t = 0$ of 5 N.

Answer (i) $\dfrac{10(1-e^{-0.5s})}{s^2}$ (ii) $\dfrac{10(1-e^{-0.5s})}{s^2(1+2s)}$ (iii) 2.91, 3.16

3 A voltage source E is connected across a resistor of 100 ohms and a 0.02 F capacitor in series. It is switched on to its value of 5 V.

(i) Derive the transfer function between i, the current through the resistor, and E.

(ii) Hence determine $i(t)$.

Answer (i) $\dfrac{0.02s}{1+2s}$ (ii) $i(t) = 0.05e^{-0.5t}$

4 A second capacitor of 0.04 F is connected in the circuit of problem **3** in series with the resistor–capacitor combination. What is the new transfer function $i(s)/E(s)$?

Answer $\dfrac{0.01s}{s+0.75}$

5 Figure 2.22 shows an *RCL* circuit with constant voltage source E. Derive the transfer function between the two voltages, $V_R(s)/E(s)$, and hence $V_R(t)$ when the voltage source is switched on to a value of 5 V. Use the values $R = 5$ ohms, $C = 0.04$ F and $L = 20$ H.

Sketch the voltage response across the resistor.

Answer $\dfrac{s}{20s^2+5s+25}$, $1.125e^{-0.125t}\sin 1.111t$

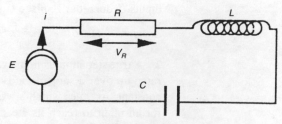

Figure 2.22 *RCL* circuit

6 A mass of 10 kg is suspended by a spring and hangs stationary in the equilibrium position with the spring extended. There is negligible damping in the system and the spring stiffness is 200 N m^{-1}. An additional mass of 2 kg is placed on the first mass. Write a differential equation for the subsequent motion of the combined masses and solve this to describe the motion as a function of time.

Answer $12\ddot{x} + 200x = 2g$, $x = 0.098(1 - \cos 4.082t)$m

7 A trailer of mass 1000 kg is attached to a vehicle using a tow hitch of spring stiffness 2×10^4 N m^{-1}. A damper in parallel exerts a damping force of 200 N m^{-1} s. The vehicle moves off at a constant acceleration of 0.7 m s^{-2}. Derive a transfer function between (a) the speed of the vehicle and the speed of the trailer and (b) the position of the vehicle and the position of the trailer.

What is the extension in the tow hitch after 2 s?

Answer $\dfrac{100+s}{5s^2+s+100}$, 60 mm

8 A disc with a moment of inertia of 5 kg m^2 rotates at 600 rpm. A braking torque of 200 N m is applied and there is a viscous friction torque coefficient of 2 N m s. How long does it take for the speed to drop to 60 rpm?

Answer 4.64 s

9 What are suitable electrical analogues for the mechanical systems in problems **7** and **8**?

10 An electric motor has a time constant T_1 and a torque to input current gain of A N m A^{-1}. It drives an inertial load with moment of inertia J kg m^2 and there is a resistive torque of $\lambda\omega$ N m where λ is a constant and ω is the speed of the load in rad s^{-1}. There is a fixed ratio gear box between the motor and load reducing the speed of the load to $1/N$ that of the motor.

(i) Draw a diagram showing the functional elements of this system.

(ii) If the current is increased by i A what is the resulting steady state increase in the speed of the load?

(iii) If there is a sudden addition to the load torque of ΔL how would this affect the speed of the load?

(iv) The output speed of the motor is measured as ω_m and compared with a set value ω_{md}. Any difference between these values gives an error $\omega_{md}-\omega_m$ which is used to adjust the motor current so that $i = K(\omega_{md}-\omega_m)$. Derive the transfer function now which relates the resulting change in speed to a change in the load of ΔL. Compare this with (iii) above.

Answer (ii) $ANi\lambda$,   (iii) reduced by $\Delta L\lambda$,   (iv) reduced by $\dfrac{\Delta L}{\lambda+A\ KN^2}$

11 For the ideal continuous stirred tank, Fig. 2.16, show that the same form of relationship occurs for a change in inlet flow temperature as for a concentration change, assuming no heat losses from the system. Derive an electrical analogue for this system.

12 In a stirred batch reactor the reaction is allowed to proceed rapidly and uncontrolled and the contents of mass M and specific heat ρ reach a final temperature θ. Cooling of the reactor contents then follows and heat is lost to water, in coils at the bottom of the reactor, which may be considered to stay at a constant temperature θ_0. The overall heat transfer coefficient is U and interface area is A. What is the system time constant and how long will it take for the initial difference in temperatures, $\theta - \theta_0$, to be reduced by half?

Answer $\dfrac{M\rho}{UA}, \dfrac{M\rho}{UA} \ln 2$

13 Elements already shown in the examples may be combined to establish models for larger overall systems and it will be necessary to select the appropriate parameters for a simplified model. (Specific computer packages may also be used in the study of larger systems to provide the computational power to deal with both linear and non-linear systems, time varying coefficients, etc.) A servo driven electrical motor is connected to a mixer, using a fluid drive for transmission of power, Fig. 2.23. Select suitable parameters and establish an overall desired speed—actual speed transfer function for the motor. (Assume that the mixer acts as a combined viscous friction plus inertial load and that the torque transmission through the coupling is proportional to the difference in speed of the input and output shafts.)

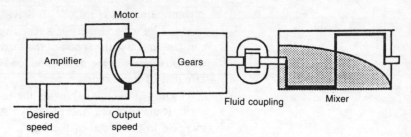

Figure 2.23 Motor driven solids mixer

3

System representation by diagrams

Benefits

It is normally an advantage in solving any problem to construct some diagrammatic representation of the data given at the start of the problem. Usually one makes some sketch of the layout in order to appreciate the physical relationship between the major components. It is of great additional advantage when this representation is capable of being manipulated to make a direct positive contribution to the solution of the problem. In the case of dynamic and control system analysis the two forms of diagram, the 'block diagram' and the 'signal flow graph', fulfil such a role.

The block diagram and signal flow graph convey information about the input/output relationship for subunits of the system, and about the flow of information or signals between these elements which make up the whole system. These elements may not coincide directly with individual physical units but will represent convenient analytical units within the overall model. Similarly, physical effects, such as friction, which act on the output of a unit, such as a driven shaft, may appear as inputs to that unit's block diagram. It is thus necessary to have an understanding of the system being considered and not to rush blindly into a direct one-to-one representation without some knowledge of the real situation. It is this need to be able to analyse the system which was emphasized when dealing with modelling and the block diagrams and signal flow graphs are themselves a form of model conveying the same information as the mathematical equations considered then.

Signal flow graphs and block diagrams have their own 'algebras' which simplify the formation of the overall relationship between the input and output of systems made up of a number of elements or functions. They are particularly useful when there are 'closed loops' within the functions of the system. Both use the Laplace transformed variables to advantage.

Signal flow graphs

Signal flow graphs consist of a number of 'nodes' representing system variables or signals and these are connected by 'branches' which are the dynamic connections between these nodes. These branches are labelled with signal gains or transfer functions. A representative graph is shown in Fig. 3.1.

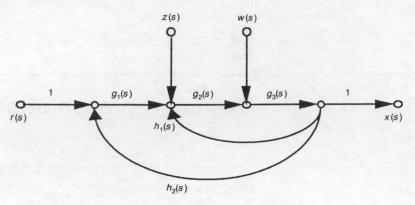

Figure 3.1 Signal flow graph

The algebra of signal flow graphs is shown in Fig. 3.2.

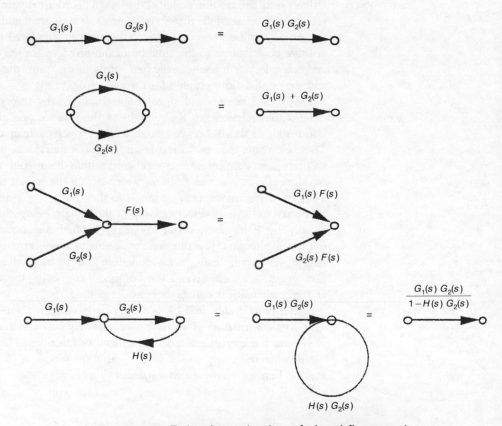

Figure 3.2 Rules for reduction of signal flow graph

As well as by use of these rules the overall transfer function between any input node i and output node j may be obtained by using Mason's rule:

$$G_{ij}(s) = \frac{\Sigma G_k(s) \Delta_k}{\Delta}$$

where

$G_k(s)$ = gain or transfer function of the kth path connecting node i to node j

Δ = determinant of graph

= 1 − sum of all different loop gains + sum of gain products of all possible combinations of two non-touching loops − sum of gain products of all possible combinations of three non-touching loops + ...

= $1 - \Sigma L_a + \Sigma L_b L_c - \Sigma L_d L_e L_f + \ldots$

and where

ΣL_a = sum of all different loop gains or transfer function

Δ_k = the term derived from Δ by keeping only those terms which are fully isolated from path k (i.e. no common branch or node), called the minor of the kth path.

3.1 First order system signal flow graph

A thermometer at an ambient temperature of 20 °C is plunged into a laboratory constant temperature bath which is at a temperature of 40 °C. If the thermometer behaves as a first order system with a time constant of 3 s draw a signal flow graph showing the input/output relationship for the thermometer.

Solution Let θ be the change in temperature of the thermometer after it is plunged into the water. The input $u(t)$ is a step change of 20 °C in its surrounding temperature. The final rise in temperature will be 20 °C and with a time constant of 3 s the equation for the change is of standard first order form

$$\theta(t) = 20(1 - e^{-t/3})$$

The corresponding transfer function, of unity steady state gain, is

$$\frac{\theta(s)}{u(s)} = \frac{1/3}{s + 1/3} \quad \text{with } u(s) = \frac{20}{s}$$

The signal flow graph is as shown in Fig. 3.3.

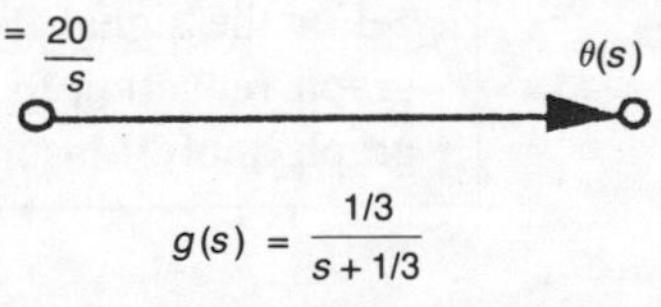

Figure 3.3 First order signal flow graph

3.2 Second order system signal flow graph

The output from a first order system is applied as input to another first order system and there is no backwards coupling between the two. If the time constants are T_1 and T_2 respectively and steady state gains are unity draw the signal flow graph representing the overall input/output relationship.

If the final output is now fed back so that the first input is the difference between a reference input and this fed back signal, show this by addition to the first diagram and reduce the full representation to a single path between two nodes.

Solution The signal flow diagram in the first case, Fig. 3.4, is comprised of just two serial elements. The transfer functions come directly from the given unity gain and the time constants.

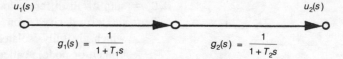

Figure 3.4 Second order signal flow graph

The addition of the feedback signal is shown in Fig. 3.5.

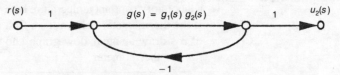

Figure 3.5 Signal flow graph with feedback

In turn this may be reduced to the single path between the input and output nodes as shown in the final diagram, Fig. 3.6.

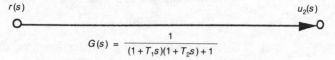

Figure 3.6 Reduced flow graph

3.3 Application of Mason's rule

For the signal flow graph shown, Fig. 3.7, use the single stages of graph reduction to derive the overall transfer function. For this same graph apply Mason's rule directly to achieve the same result.

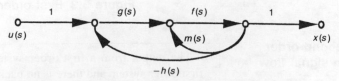

Figure 3.7 Flow graph with nested feedback

Solution The progressive stages of reduction, starting from the given signal diagram, are shown in the diagrams, Fig. 3.8(*a* to *c*).

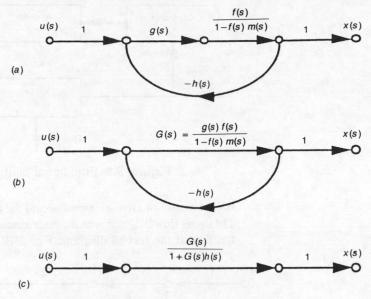

(a)

(b)

(c)

Figure 3.8 Nested flow graph reduction

From the final diagram the overall transfer function between the input node and the output node is

$$\frac{x(s)}{u(s)} = \frac{G(s)}{1+G(s)h(s)}$$

$$= \frac{g(s)f(s)}{[1-f(s)m(s)] + g(s)f(s)h(s)}$$

The same overall transfer function should naturally be given using Mason's rule. Applying this directly to the given diagram it is seen that there is just the one path connecting the two end nodes, there are two loops with loop gains $f(s)m(s)$ and $-g(s)f(s)h(s)$ respectively and no further terms. The transfer function may be written directly as above.

3.4 Signal flow graph for given functional diagram

Draw the signal flow graph for the system shown in Fig. 3.9 in which the vessel level is controlled by the flow into it through an electrically driven valve. Hence derive the change in level following a step change in level demand. Assume there is no outflow, i.e. $d = 0$.

Solution Assume that the vessel dynamics dominate the behaviour, with the other elements of the system being represented solely by fixed gains. The controller operates with gain k_d on the error signal, i.e. controller output is $k_d(r-h)$. The series gains k_a, k_m, and k_v may be simply multiplied together.

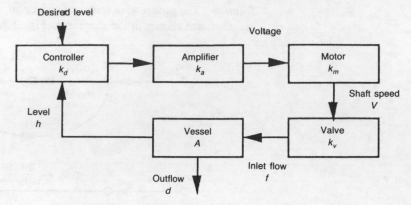

Figure 3.9 Functional diagram for level control

For the vessel of cross-sectional area A the dynamic equation is $dh/dt = f/A$. The signal flow diagram is as shown, reducing to the first order overall transfer function of the second diagram, Fig. 3.10.

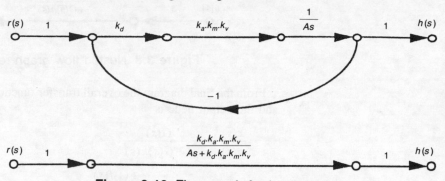

Figure 3.10 Flow graph for level control

For a step change of magnitude R in the required height, $r(s) = R/s$,

$$h(s) = \frac{k_d k_a k_m k_v}{(As + k_d k_a k_m k_v)} \frac{R}{s}$$

and

$$h(t) = R(1 - e^{-kt/A})$$

where $K = k_d k_a k_m k_v$.

Block diagrams

Although the block diagram is usually drawn incorporating transformed variables it may also be shown in terms of the variables as functions of time, e.g. $x(t)$, rather than of the complex variable, $x(s)$. The impulse function $g(t)$ then appears within the block. If the transfer function representation is valid then the input and output are given as transformed variables and the block contains the transfer function, Fig. 3.11.

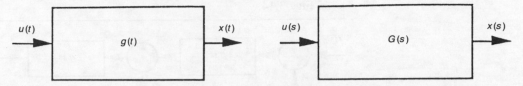

Figure 3.11 Fundamental block diagram

As with the signal flow graphs, block diagrams enable much of the simplification of system representation to be carried out rapidly without recourse to the algebraic elimination of intermediate variables etc. in the equation form, e.g. Fig. 3.12.

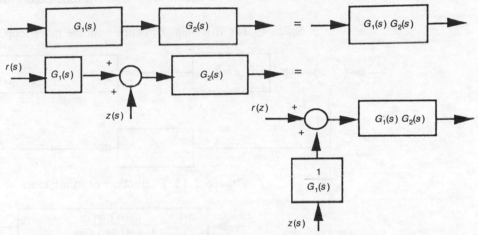

Figure 3.12 Basic block diagram algebra

For a system with a feedback loop the representation by a single block is a critical operation, simplifying the reduction of the system representation in many cases, Fig. 3.13.

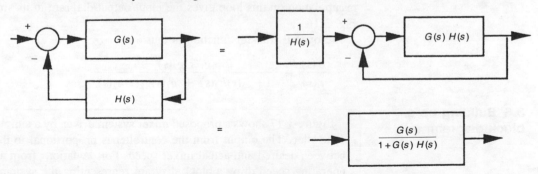

Figure 3.13 Block diagram algebra for negative feedback loop

3.5 Block diagram reduction

Using the standard block diagram reduction rules derive the overall transfer function for the system represented by the block diagram given in Fig. 3.14.

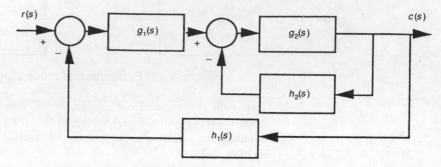

Figure 3.14 System block diagram

Solution The first step is to cope with the inner loop, Fig. 3.15.

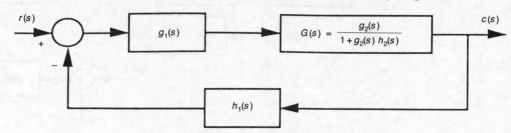

Figure 3.15 Reduction of inner loop

$$\frac{g_1(s)\, G(s)}{1 + g_1(s)\, G(s)\, h_1(s)}$$

Figure 3.16 Final reduced form

This has reduced the diagram to a representation with just one negative feedback loop. Combining the two forward path blocks and repeating the above manipulation on this loop gives the input/output diagram in its simple form, Fig. 3.16.

The overall transfer function is thus

$$\frac{c(s)}{r(s)} = \frac{g_1(s)g_2(s)}{1 + g_2(s)h_2(s) + g_1(s)g_2(s)h_2(s)}$$

3.6 Building up a block diagram

Figure 3.17 shows a proposed mixer system driven by a simple speed controller. The output from the controller is proportional to the error between desired and actual mixer speed. For deviations from a steady operating speed draw a block diagram representing the system, using separate blocks for each element of the system function.

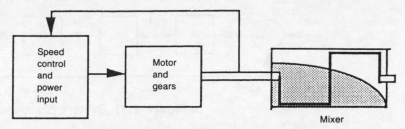

Figure 3.17 Mixer with speed control

Solution Before a start is made on the block diagram the system itself needs to be analysed. In this case the control function is that the output u is proportional to the error between the desired speed of the mixer ω_r and the actual speed ω_m. Assume that the motor torque T is proportional (by factor k) to the control output u. The torque at input to the mixer will be NT where N is the step down gear ratio between the motor and the input shaft to the coupling. If the speed of the motor is ω_d the speed of this shaft is ω_d/N.

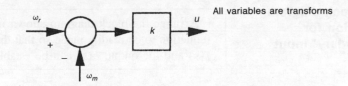

Figure 3.18 Controller representation

Figure 3.19 Motor and gears block diagrams

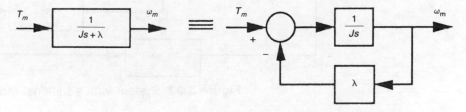

Figure 3.20 Alternative mixer block diagrams

Moving to the mixer, assume that it may be represented by an inertial load J and viscous friction λ and turns at a speed ω_m. The applied torque is NT.

The block diagram contributions for the full system are as shown in the individual figures, Figs 3.18, 3.19 and 3.20.

These may be combined to give the full block diagram, Fig. 3.21.

The alternative representation of the mixer to show the effective reduction of the input torque, by the viscous drag, to give the net accelerating torque leads to the alternative diagram in Fig. 3.22.

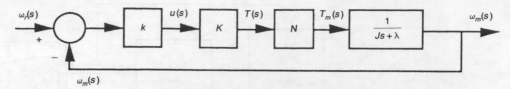

Figure 3.21 System block diagram

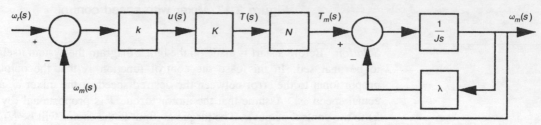

Figure 3.22 Block diagram showing feedback
configuration for friction

3.7 Block diagram reduction for 'secondary' input

Arrange the block diagram shown in Fig. 3.23 in a form to enable reduction to be readily made so that the relationship between the input $z(s)$ and the output $c(s)$ can be established.

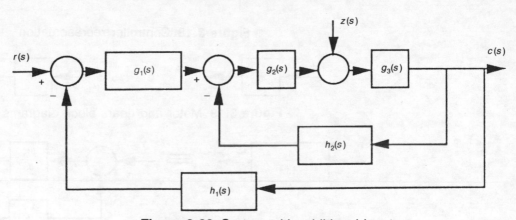

Figure 3.23 System with additional input

Solution In general the block diagram reduction and manipulation rules are most easily recalled and used in the 'input at left' and 'output at right' form shown so far. The rearrangement and reduction are shown in the following stages. As interest is centred on the input z the second input r may be set to zero and does not appear in the following diagrams. Because of the superposition property separate solutions for the two inputs may be added together if required.

First realign the input, Fig. 3.24.

The following equivalent, Fig. 3.25, makes reduction of the loops easier. Remove the inner loop first, Fig. 3.26.

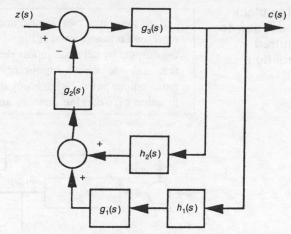

Figure 3.24 Realignment of diagram to place input $z(s)$ conveniently

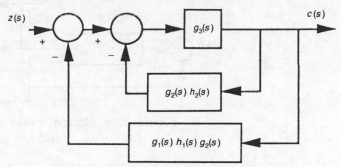

Figure 3.25 Standard nesting of feedback loops

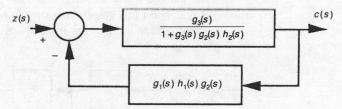

Figure 3.26 Reduction of inner loop

Final reduction by removing the second feedback loop yields eventually Fig. 3.27.

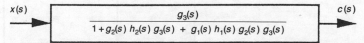

Figure 3.27 Final diagram with single overall transfer function

Note that blocks $g_1(s)$ and $g_2(s)$ which appear in the original forward path from $r(s)$ to $c(s)$ are in the feedback path of the z to c relationship.

3.8 Complex block diagram reduction to form a required transfer function

A system has two inputs and two outputs with a block diagram representation as shown in Fig. 3.28. Although the intention is to control the outputs by separate inputs the interaction in the system complicates this and the direct input/output relationships between the chosen input/output pairs. Using block diagram reduction establish the transfer function between the input r_1 and the output c_2.

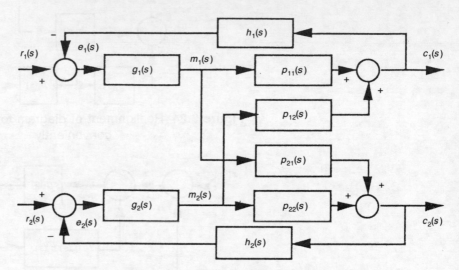

Figure 3.28 More complex system having two inputs and two outputs

Solution As the relationship between r_1 and c_2 is required, show this as the main path in the block diagram. This gives the diagram in Fig. 3.29.

Once again this can be more usefully drawn as shown in Fig. 3.30.

Now reduce each of the inner loops, Fig. 3.31.

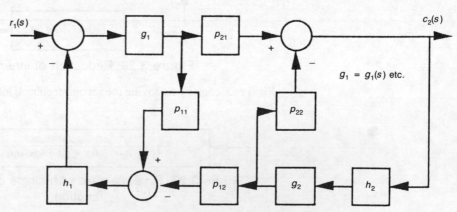

$g_1 = g_1(s)$ etc.

Figure 3.29 Diagram for required input/output relationship

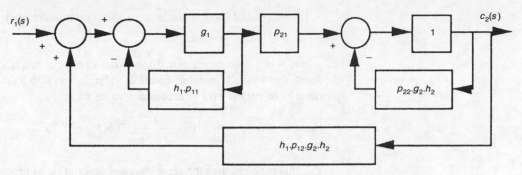

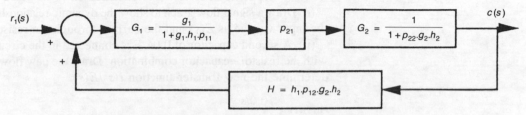

Figure 3.30 Alternative diagram clarifying nested loops

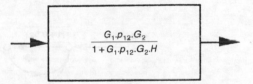

Figure 3.31 Diagram after reduction of inner loops

Finally the overall transfer function is given in terms of the composite functions, Fig. 3.32.

$$\frac{G_1.p_{12}.G_2}{1 + G_1.p_{12}.G_2.H}$$

Figure 3.32 Final block diagram and required transfer function

Although algebraic manipulation could be used as an alternative, this block diagram approach divides the search for the relationship into a number of simpler stages where very little algebra, more prone to errors, is required. Such reductions are also available as elements of computer packages.

Problems

1 A first order dynamic system is represented by the differential equation

$$5\frac{dc}{dt} + c = u(t)$$

(i) Derive the transfer function and show it on a signal flow graph.

(ii) A control loop is formed for this system such that

$$u = r - c$$

where r is a reference input. Extend the first flow graph to show this, form the overall transfer function relating $c(s)$ to $r(s)$ and hence determine $c(t)$ if $r(t)$ is a constant value of unity.

Answer   (i) $\dfrac{1}{1+5s}$,   (ii) $\dfrac{1}{2+5s}$, $0.5(1-e^{-0.4t})$

2   This problem uses the same physical description as problems **2.3** and **2.4**. A voltage source E is connected across a resistor of 100 ohms and a 0.02 F capacitor in series. It is switched on to its value of 5 V.
(i) Draw a signal flow graph to show the dynamic relationship between the input voltage as $E(s)$ and the current through the resistor as $i(s)$.
(ii) A second capacitor of 0.04 F is connected in the circuit in series with the resistor—capacitor combination. Draw the new flow graph and determine the new transfer function $i(s)/E(s)$.

Answer $\dfrac{0.0/s}{s+0.75}$

3   Reduce the signal flow graphs in Fig. 3.33 to single transfer functions between the input variables $r_1(t)$ and $r_2(t)$ and the output $c(t)$.

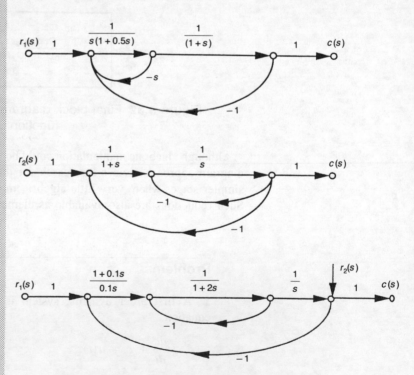

Figure 3.33 Signal flow graphs for reduction

Answer (i) $\dfrac{1}{s(s+1)(2+0.5s)+1}$ (ii) $\dfrac{1}{(s+1)^2+1}$

(iii) $\dfrac{1+0.1s}{0.2s^2(1+s)+0.1s+1}$, $\dfrac{0.2s^2(s+1)}{0.2s^2(s+1)+0.1s+1}$

4 For the first two flow graphs in Fig. 3.33 use Mason's rule to check the corresponding transfer functions.

5 For the system in Fig. 2.17 draw a signal flow graph when the second tank is full and has a constant exit flow q. Establish the transfer function between the solids input $u(t)$ and the concentration in the second tank $c_2(t)$.

Answer $\dfrac{q}{(q+V_1s)(q+V_2s)}$

6 Figure 3.34 shows a block diagram representing a mechanical system which is subjected to two distinct inputs. Why can the combined effects of these inputs be obtained by considering each in turn in isolation?

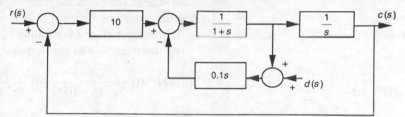

Figure 3.34 System with two inputs

By suitable block diagram manipulation derive (i) the transfer function between $r(t)$ and $c(t)$, and (ii) the transfer function between $d(t)$ and $c(t)$.

Answer (i) $\dfrac{10}{s(1+1.1s)+10}$ (ii) $\dfrac{-0.1s}{s(1+1.1s)+10}$

7 A second order device has the following transfer function between the input x and the output y:

$$\frac{y(s)}{x(s)} = \frac{10}{s(s+10)}$$

The input is controlled by a controller with simple proportional gain K between its input e and output m, i.e.

$$m(s) = Ke(s)$$

The error is $e = r-y$, where r is the reference input to the controller. The output from this control term is supplemented by the addition of 'velocity feedback' so that

$$x(s) = m(s)-0.2s.y(s)$$

(i) Show these relationships on a block diagram and determine the overall transfer function $y(s)/r(s)$.

(ii) If $K = 1$ determine $y(t)$ if $r(t) = \mathcal{H}(t)$, i.e. a unit step.

Answer (i) $\dfrac{10K}{s^2+12s+10K}$ (ii) $y(t) = 1-1.089e^{-0.9t}+0.089e^{-11.1t}$

8 A trailer of mass 1000 kg is attached to a vehicle using a tow hitch of spring stiffness 2×10^4 N m^{-1}, problem **2.7**. A damper in parallel exerts a damping force of 200 N m^{-1} s.

(i) Draw a block diagram and from it derive a transfer function between the position of the vehicle and the position of the trailer as the vehicle moves away from rest. (It may be found beneficial to start in terms of the two velocities.)

(ii) If the trailer is itself attached to a fixed post by an elastic rope of stiffness 1×10^5 N m^{-1} show that this adds a feedback loop to the block diagram. What are the limitations to this analysis?

9 Draw the equivalent block diagram for the fourth flow graph in Fig. 3.33. Using the block diagram reduction algebra, confirm the overall transfer functions between

(i) the input r_1 and the output c, and

(ii) the input r_2 and the output c.

Answer (i) $\dfrac{100+s}{5s^2+s+100}$, (ii) $\dfrac{100+s}{5s^2+s+600}$

10 A d.c. motor is represented by a first order transfer function of steady state gain K and time constant T_m. The output torque is used to drive a mechanism which moves a mass m along a slide against the action of a spring of stiffness f. The torque to linear force conversion is given by a constant gain term K_l. The slide is not friction free but exerts a viscous friction force opposing motion with coefficient λ, i.e. frictional force magnitude is λ times the velocity of the mass. The position of the mass is controlled by measuring the position, the difference between this and the preset required value giving an error value. A controller which has its own dynamics and transfer function $K_c(s)$ converts this error to the current input to the motor.

(i) Write out the governing differential equations and transfer functions for each of the system components.

(ii) Draw a block diagram and by block diagram reduction determine the overall transfer function of this closed loop system.

(iii) Compare this method with pure algebraic manipulation.

Answer (ii) $K_c(s)KK_l/[(ms^2+\lambda s+f)(1+T_m s)+K_c(s)KK_l]$

11 Process systems involving a number of vessels, reaction stages, flow in pipes, mixing, etc., lead to complex models which may sometimes be reduced to quite simple stage representations as the actual transients for the stages approximate to first or second order lags and/or pure delays. An illustrative form of block diagram has been given in Fig. 3.28 in Example 3.8. This may be simplified by assigning the following specific values to the individual transfer functions:

$$h_1(s) = 1 \qquad\qquad h_2(s) = 1$$
$$g_1(s) = 2 \qquad\qquad g_2(s) = 3$$

$$p_{11}(s) = \frac{1}{1+0.1s} \qquad p_{12}(s) = \frac{0.1}{1+0.5s}$$

$$p_{21}(s) = 0 \qquad\qquad p_{22}(s) = \frac{1}{1+0.3s}$$

Consider now that noise enters the system in the lower feedback loop, i.e. an input $n_2(s)$ is added to the feedback of $c_2(s)$ immediately prior to $h_2(s)$. Establish by block diagram manipulation (i) the transfer function $c_2(s)/n_2(s)$ and (ii) the transfer function $c_1(s)/n_2(s)$.

Answer　(i) $\dfrac{-3}{4+0.3s}$　(ii) $\dfrac{-0.3(1+0.1s)(1+0.3s)}{(1+0.5s)(3+0.1s)(4+0.3s)}$

4

Systems with simple feedback

Feedback

Feedback within a system can be present through the physics of that system, e.g. the restraining force introduced by viscous friction, or it can be deliberately introduced in order to stabilize or improve the dynamic or steady state behaviour. This may be effected through the physical addition of components, such as dashpots, or through the addition of 'feedback control' loops. If control action is made without the benefit of this feedback then the system will be under 'open loop' or 'feedforward' control. Although this may be applicable at times it is unable to respond to output deviation resulting from poor information about the system or from unmeasured disturbances. Feedback overcomes the problems associated with such lack of information of the system or inputs.

Simple feedback control uses a controlling input which is proportional to the error or difference between the desired output and the measured output. This is 'proportional' control with 'negative feedback'. The extension of this form of control by the addition of extra terms, considered more generally later, leads to controllers having greater potential for defining more closely the 'closed loop' system behaviour. The function of control is initially to establish, where necessary, the 'stability' of the system and secondly to improve steady state and dynamic performance.

The general block diagram for the initial study of feedback systems is given in Fig. 4.1.

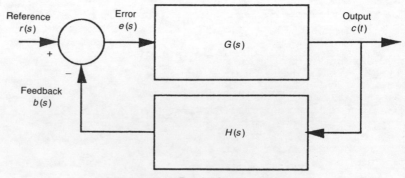

Figure 4.1 Basic feedback loop

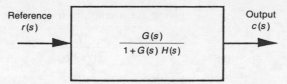

Reference
$r(s)$ | $\dfrac{G(s)}{1+G(s)\,H(s)}$ | Output
$c(s)$

Figure 4.2 Single block equivalent of feedback loop

Note that this is readily reduced to the single block representation, Fig. 4.2, so that all of the techniques applicable to looking at the dynamics of simple systems without feedback remain applicable here.

The forward path transfer function $G(s)$ will in general be comprised of a constant gain term K and a term representing the system open loop dynamics in the absence of feedback, say $G_1(s)$, such that

$$KG_1(s) = G(s)$$

Both $G_1(s)$ and $H(s)$ may be made up of the transfer functions of a number of system elements. The effect of simple proportional control action is to close the feedback loop and use changes in the value of the gain K to give, as far as possible, the system's desired performance.

4.1 Damping as negative feedback

Natural feedback occurs in some systems, as is demonstrated by the addition of viscous friction to an otherwise undamped mechanical inertial system. A shaft and mass of moment of inertia J is driven by a constant torque T. Show that in the absence of friction the velocity ω will continue to increase with time but that the introduction of a viscous damper adds feedback to the system which stabilizes the motion.

Solution In the absence of friction the equation of the system dynamics is

$$T = J \frac{d\omega}{dt}$$

For a constant torque $T(s) = T/s$ and from zero initial conditions the governing transform equation is thus

$$\omega(s) = \frac{T}{Js^2}$$

to give

$$\omega(t) = \frac{Tt}{J}$$

i.e. steady continuous increase in speed.

With the addition of a viscous drag torque $b\omega$ the dynamic equation becomes

$$T = J \frac{d\omega}{dt} + b\omega$$

giving for a constant torque T

$$\omega(s) = \frac{T}{s(Js+b)}$$

leading to

$$\omega(t) = \frac{T}{b}(1-e^{-bt/J})$$

The speed now reaches a limiting, stable, value J/B which depends on the frictional torque. The feedback effect of this friction is illustrated by looking at suitable block diagrams for the two cases, Fig. 4.3.

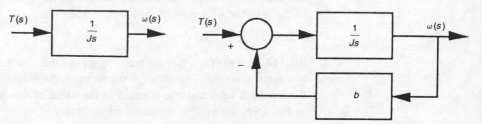

Figure 4.3 Torque to speed relationships without and with friction

4.2 Effect of adding feedback to brake loaded inertia

If a system is open loop stable, i.e. its response to an input is bounded, then it may be 'controlled' by changing its input to give a change in the output. This open loop control simply utilizes the dynamic behaviour of the system. However, it is unable to cope with changes in the system or deviations caused by other inputs. Figure 4.4 shows an inertial load driven through a viscous drive from an electric motor.

(i) If the motor is run at constant speed and the friction brake torque is doubled what is the effect on the output shaft speed?

(ii) If the motor speed is set by a controller so that for a drop in output speed of 1 rad s^{-1} the motor speed changes by k rad s^{-1} how does this improve the behaviour of the system following the same change in braking torque?

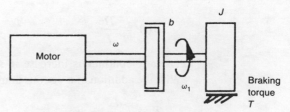

Figure 4.4 Open loop braked system

Solution   (i) Torque transmitted to load $= b(\omega - \omega_1)$

$$= J\frac{d\omega_1}{dt} + T$$

At steady conditions $d\omega_1/dt = 0$ and $b(\omega - \omega_1) = T_0$ where T_0 is the steady brake torque. If the motor speed ω is constant but the brake torque is now changed to $2T_0$ the dynamics equation becomes

$$b[\omega - (\omega_1 + \delta\omega_1)] = J\frac{d(\omega_1 + \delta\omega_1)}{dt} + 2T_0$$

where $\delta\omega_1$ is the change in speed. Subtracting the initial conditions gives

$$-b\,\delta\omega_1 = J\frac{d\,\delta\omega_1}{dt} + T_0$$

T_0 is a constant step so

$$\delta\omega_1(s) = \frac{-T_0}{s(Js+b)}$$

and

$$\delta\omega_1(t) = \frac{-T_0}{b}(1 - e^{-bt/J})$$

(ii) If the output shaft speed is now linked by the controller to the motor speed then the motor speed will also change from ω to $\omega - k\delta\omega_1$. The equation following a change becomes

$$b[(\omega - k\delta\omega_1) - (\omega_1 + \delta\omega_1)] = J\frac{d(\omega_1 + \delta\omega_1)}{dt} + 2T_0$$

and the new equation in $\delta\omega_1$ is

$$-bk\,\delta\omega_1 - b\,\delta\omega_1 = J\frac{d\,\delta\omega_1}{dt} + T_0$$

Collecting terms as before leads to

$$\delta\omega_1(s) = \frac{-T_0}{s[Js+b(1+k)]}$$

which on inversion gives

$$\delta\omega_1(t) = \frac{-T_0}{b(1+k)}(1 - e^{-b(1+k)t/J})$$

Thus the feedback has reduced the drop in speed resulting from the extra braking and there is a corresponding reduction in the time constant, the new steady state speed being approached more rapidly.

4.3 Deliberate addition of damping to improve behaviour

A freely revolving rotating door is placed in an opening in a plain wall. It creates problems in that it continues to move after a person has passed through it. To overcome this it has a shield built round it as shown in Fig. 4.5 which creates a heavy damping of 500 N m rad s^{-1}. If the door has a moment of inertia of 100 kg m^2 how does the motion decay if it is left at an initial velocity of 0.7 rad s^{-1} after use?

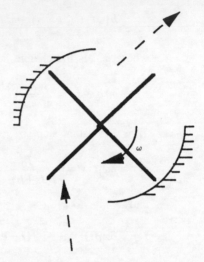

Figure 4.5 Revolving door system

Solution The physical torque applied by the door surround through the velocity dependent damping is a feedback mechanism. It depends on the door velocity, i.e. on the state of the system. There is no other force considered in the model of the behaviour and the movement of the door depends on its initial velocity. The equation of motion is

$$J \frac{d\omega}{dt} + b\omega = 0$$

Using the initial condition $\omega(0)$ and taking transforms gives

$$J[s\omega(s) - \omega(0)] + b\omega(s) = 0$$

i.e.

$$\omega(s) = \frac{\omega(0)}{s + b/J}$$

$$\omega(t) = \omega(0)e^{-bt/J}$$

The damping acts as a negative feedback loop. Although this is an exponentially decreasing function other (unmodelled) friction forces ('stiction') will act to bring the door to rest in a finite time.

4.4 Assessment of effect of controller gain on system response

A system with unity negative feedback is shown by its block diagram in Fig. 4.6. Determine qualitatively (i.e. without full solution) how this system reacts to a unit impulse as the controller gain increases from $k = 1$ to $k = 7$.

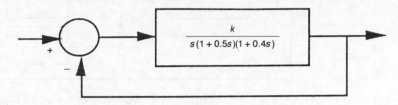

Figure 4.6 Feedback with variable gain

Solution The qualitative nature of the system response is determined by evaluating the system poles, the roots of the characteristic equation. If required, full evaluation of the response to a given input may then be made by use of the standard forms of solution. For the given system the closed loop standard form transfer function from input to output is

$$\text{CLTF} = \frac{G(s)}{1 + G(s)H(s)}$$

$$= \frac{k}{s(1 + 0.5s)(1 + 0.4s) + k}$$

with characteristic equation

$$s(1 + 0.5s)(1 + 0.4s) + k = 0$$

For $k = 1$ the roots of this equation are determined to be

$$s = -3.475, \ s = -0.051\,25 + j0.1085, \ s = -0.051\,25 - j0.1085$$

Thus all of the roots have negative real parts. The complex pair lead to an oscillatory decaying component and the real root to an exponentially decaying term.

For $k = 7$ the roots of this equation are determined to be

$$s = -4.926, \ s = 0.2138 + j0.2657, \ s = 0.2138 - j0.2657$$

Although there is still a rapidly decaying component the response will be dominated by the complex pair of roots, now having positive real parts, which give growing oscillations of the output of the system.

Note that the beneficial effect of feedback depends on the gains in the system. Feedback is also capable of producing less desirable effects in the system behaviour.

4.5 Use of velocity feedback

The closed loop system shown in Fig. 4.7 has sustained oscillations if disturbed from its equilibrium condition. Show that feedback of the velocity value can be used to reduce these oscillations.

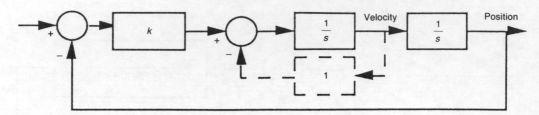

Figure 4.7 Addition of velocity feedback

Solution Without the velocity feedback the forward transfer function is k/s^2 and the closed loop transfer function (CLTF) is

$$\text{CLTF} = \frac{k}{s^2+k}$$

The roots of this are $s = j\sqrt{k}$ and $s = -j\sqrt{k}$ and lead on disturbance of the system, e.g. by an impulse at the input, to undamped oscillations.

With the added velocity feedback the first integrator term and the inner loop reduce to

$$\frac{1/s}{1+1/s} = \frac{1}{1+s}$$

and overall

$$\text{CLTF} = \frac{k}{s(1+s)+k}$$

This has the characteristic equation roots

$$s = \frac{-1 \pm \sqrt{(1-4k)}}{2}$$

giving a stable *decaying* response to the impulse input. This will be oscillatory or not depending on the magnitude of k but will still diminish as time increases.

4.6 Closed loop transfer function via signal flow diagram

A physical system is controlled using feedback of the output velocity and position. The block diagram representation is as shown in Fig. 4.8. The velocity feedback component in the control action and the gain K may be varied.

(i) Investigate, by finding the roots of the closed loop characteristic equation, the response of the system for $K = 6$ with $k_v = 11$ and then with the velocity feedback removed, i.e. $k_v = 0$.

(ii) Reduce the full diagram to the input/output transfer function using the equivalent signal flow diagram.

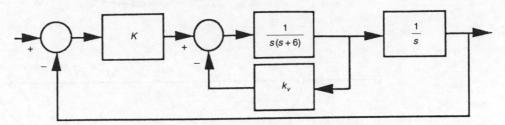

Figure 4.8 Stabilizing with velocity feedback

Solution (i) This system is third order. With $k_v = 11$ the inner loop reduces to

$$\frac{1}{s(s+6)+11}$$

and the forward transfer function with $K = 6$ to

$$\frac{6}{s[s(s+6)+11]}$$

The CLTF becomes

$$\text{CLTF} = \frac{6}{s^3+6s^2+11s+6}$$

$$= \frac{6}{(s+1)(s+2)(s+3)}$$

having poles at -1, -2, -3 and giving a damped non-oscillatory response.

Removal of the velocity feedback inner loop gives

$$\text{CLTF} = \frac{6}{s^3+6s^2+6}$$

which now has poles at -6.1582, $0.0791 + j1.010$ and $0.0791 - j1.010$. The new system is thus unstable with $K = 6$ and will also be for all positive values of K.

(ii) The signal flow graph reduction is shown in Fig. 4.9.

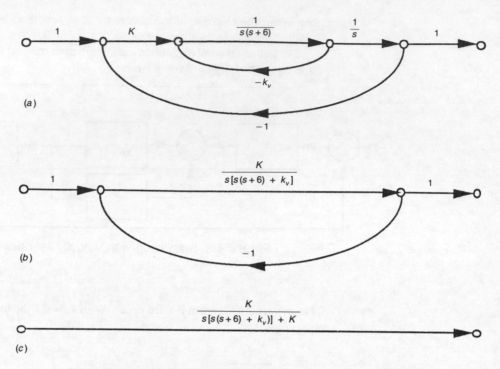

(a)

(b)

(c)

Figure 4.9 Signal flow graph including velocity feedback

Problems

1 A ram of mass M is driven by a force $f(t)$ along a guideway. It is resisted by a spring force proportional (with factor k) to the distance moved by the mass. Express the dynamics as transformed variables and draw simple alternative block diagrams or signal flow graphs for these. How does the position x of the mass change with time if $f(t)$ is a small impulse $F\delta(t)$?

If viscous friction becomes a significant factor so that the movement is resisted by an additional force $\lambda.dx/dt$ add this to the block diagram. By evaluating the impulse response of the system now demonstrate the effect of this viscous damping term in changing the system behaviour.

Answer $\quad \dfrac{F}{\sqrt{(km)}} \sin \omega t, \quad \omega = \sqrt{(k/m)}$

$$\frac{F}{\sqrt{(km - \lambda^2/4)}} \, e^{-\lambda t/2m} \, \sin[\sqrt{(k/m - \lambda^2/4m^2)}t]$$

2 A mechanical support is represented as a mass−spring−damper (M, k, λ) combination arranged with the mass supported on a spring with the damper in parallel. Disturbances may be present as vertical forces f on the mass or as movement y at the base of the spring−damper support.

Discuss the role of the damper in the system and derive the mathematical equations for the movement of the mass in the presence of each disturbance.

Answer $M\ddot{x}+\lambda\dot{x}+kx = f(t)$, $M\ddot{x}+\lambda\dot{x}+kx = \lambda\dot{y}+ky$

3 A plant has the transfer function $g(s) = 1/[s(1+1.2s+s^2)]$. It is controlled by a proportional controller of variable gain K with unity negative feedback.

(i) Show that the value of $K(>0)$ has no influence on the stable closed loop steady state value following a change in demand.

(ii) What is the effect of using a negative value of K?

(iii) By evaluation of the roots of the characteristic equation describe the system behaviour for $K = 1.0, 1.2, 1.5$.

Answer   (ii) Instability, (iii) Stable, limiting stability, unstable

4 A controller of gain K is in series in a negative feedback loop with a plant having the transfer function $1/[s(1+0.05s)]$.

(i) If the feedback loop is open how does the system respond to (a) a unit impulse and (b) a unit step input?

(ii) The loop is now closed. How does the system react to a unit step input if $K = 0.5$?

Answer   (i) $K(1-e^{-20t})$, $K(t+0.05e^{-20t}-0.05)$

(ii) $1+0.027e^{-19.49t}-1.027e^{-0.51t}$

5 (i) Draw a block diagram or signal flow graph to show the addition of unity negative feedback to a system with a combined controller and plant forward path transfer function $G(s)$.

(ii) If $G(s)$ is given by

$$G(s) = \frac{K}{(1+s)(1+0.5s)}$$

what value of K (proportional controller gain) will give a steady state error of just 1%?

(iii) With this value of K the response is oscillatory and very lightly damped. Velocity feedback will improve the dynamic response and improve the damping. (The principle of velocity feedback is shown in Fig. 4.7.) What value of velocity feedback gain k_v will give a closed loop damping coefficient of 0.8?

Answer   (ii) 99, (iii) 9.81

6 The analysis of feedback systems containing a pure delay T suffers from the introduction of the term e^{-Ts} in the transfer function. For example, the use of certain instrumentation may introduce a pure delay in the feedback loop or a pure delay may occur in the plant or process itself.

A heat exchange device is approximately modelled by the transfer function

$$g(s) = \frac{e^{-s}}{1 + 10s}$$

By considering this in a closed loop with a controller $K(s)$ discuss the difficulty of producing an analytical solution to the system with feedback.

7 Although frictional forces can add stabilizing effects, as with the normal damping dashpot or frictional brake, they can also have the reverse effect. An example of this has been shown to be the force produced on a wire, e.g. by wind on power transmission lines, under certain conditions. A transverse force is set up in the same direction as the instantaneous deflection velocity of the wire. This is in contrast to the more usual effect of viscous forces which act so as to oppose the instantaneous velocity. Consider the forces on the wire shown in Fig. 4.10 and show that there is a feedback effect which in this case causes oscillations to grow with time, i.e. there is positive feedback.

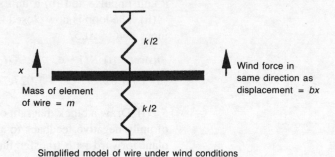

Simplified model of wire under wind conditions

Figure 4.10 Forced mechanical system

5

System poles, zeros and stability

Rational transfer function

Assume that a transfer function $G(s)$ expressing the relationship between the input and output of a system may be written as the ratio of two polynomials. Thus

$$G(s) = \frac{p(s)}{q(s)}$$

$$= \frac{k(s-z_1)(s-z_2) \ldots (s-z_m)}{(s-p_1)(s-p_2) \ldots (s-p_n)}$$

where the roots of the numerator are the 'zeros' of the system transfer function (i.e. of the system in the absence of any internal pole−zero cancellation) and the roots of the denominator are the 'poles' of the transfer function similarly.

System poles

The poles of the transfer function are the roots of the 'characteristic equation',

$$q(s) = 0$$

The positions of these poles in the complex plane determine the stability of the system. This is of particular significance when considering the effect of feedback on a system, which may be open loop stable or unstable, and of the variation of controller gains. The transfer function may be expanded in terms of its poles so that

$$G(s) = \frac{a_1}{s-p_1} + \frac{a_2}{s-p_2} + \ldots + \frac{a_n}{s-p_n}$$

If an open loop or closed loop system, with transfer function $G(s)$, is subject to a unit impulse input, for which $U(s) = 1$, then the output, say $x(t)$, is given by the inversion of $G(s)$, i.e.

$$x(t) = a_1 e^{p_1 t} + a_2 e^{p_2 t} + \ldots + a_n e^{p_n t}$$

This is the system impulse response function, frequently given the specific

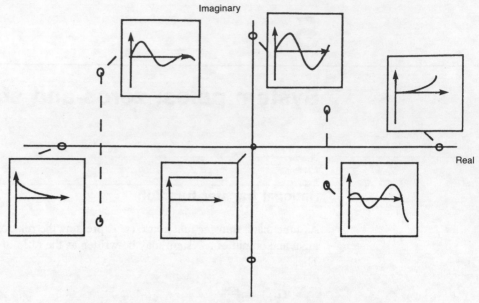

Figure 5.1 First order system behaviour with respect to
pole position, response to impulse input

notation $g(t)$. Provided the real part of all p_i is negative then the roots of the characteristic equation, the system poles, lie in the left half of the complex plane and the system response to a bounded input will always be bounded and decay at large times, Fig. 5.1. A pair of poles on the imaginary axis, i.e. real part of the poles equal to zero, will lead to constant amplitude oscillations of the system output (marginally stable), and if the poles fall in the right half of the plane then the response will grow unbounded with time and the system is unstable.

Stability

As indicated above, the stability of both closed and open loop systems is determined by the positions of the system poles. For linear systems the stability is global in that the fundamental behaviour of the system is not affected by the operating point of the system nor by the magnitude of the input. In real physical systems this is not usually the case, in that conditions such as saturation or other physical limitations break the requirements for superposition and linearity. A system is said to be stable if, for a bounded input, the output remains bounded, i.e. it does not grow or oscillate with ever increasing magnitude. If the output oscillates at constant amplitude after the decay of other transient effects then the system is said to be 'marginally stable', with small differences in gain or other parameters possibly causing full instability.

   A fuller treatment of stability and the criteria for assessing it are given in Chapters 8 and 9.

5.1 Stable and unstable poles and system response

By considering the pole location for the systems described by the given differential equations comment on the stability of these systems. By evaluating the response to a unit step input illustrate that for a bounded input the output of the system is also bounded if the system has all stable poles.

(i) $\quad \dfrac{d^2x}{dt^2} + 3\dfrac{dx}{dt} + 2x = u(t)$

(ii) $\quad \dfrac{d^2x}{dt^2} + \dfrac{dx}{dt} - 6x = u(t)$

(iii) $\quad \dfrac{d^3x}{dt^3} + 5\dfrac{d^2x}{dt^2} + 8.25\dfrac{dx}{dt} + 4.25x = u(t)$

Solution Consider each system equation in turn.

(i) Expressing the equation in transformed variables,

$$(s^2+3s+2)x(s) = u(s)$$

The input/output transfer function is

$$\frac{x(s)}{u(s)} = \frac{1}{s^2+3s+2} = \frac{1}{(s+1)(s+2)}$$

Thus there are system poles at $s = -1$ and $s = -2$ which give rise to stable and non-oscillatory responses. Confirmation is given by the response to a unit step input when $u(s) - 1/s$ and

$$x(s) = \frac{1}{s(s+1)(s+2)} = \frac{0.5}{s} - \frac{1}{s+1} + \frac{0.5}{s+2}$$

to give

$$x(t) = \mathbf{0.5} - \mathbf{e}^{-t} + \mathbf{0.5e}^{-2t}$$

(ii) In this case

$$\frac{x(s)}{u(s)} = \frac{1}{(s+3)(s-2)}$$

giving one stable and one unstable pole (at -3 and $+2$ respectively) thus giving overall instability to the system. For the same step input the system response is

$$x(t) = -\frac{1}{6} + \frac{1}{15}\,\mathbf{e}^{-3t} + \frac{1}{10}\,\mathbf{e}^{2t}$$

with the exponentially growing term coming from the 'unstable' pole.

(iii) This third order relationship factorizes to give first and second order factors,

$$\frac{x(s)}{u(s)} = \frac{1}{(s+1)(s^2+4s+4.25)}$$

having poles at -1 and $-2 \pm j0.5$. These are all stable giving a decaying response with oscillations for the disturbed system. With a unit step input again

$$x(s) = \frac{0.2353}{s} - \frac{0.8}{s+1} + \frac{0.5647s + 1.447}{(s+2)^2 + 0.25}$$

and

$$x(t) = \mathbf{0.2353 - 0.8e^{-t} + 0.5647e^{-2t}\cos 0.5t + 0.6347e^{-2t}\sin 0.5t}$$

5.2 Effect of open loop gain on closed loop poles

> An open loop system has the transfer function
>
> $$\frac{x(s)}{u(s)} = \frac{K}{(1+0.5s)(1+s)}$$
>
> What is the transfer function when applying unity negative feedback to this system? Discuss the closed loop pole movement and system behaviour as the gain K is increased, using values for K of 0.1, 0.5, 4.

Solution For this system the closed loop transfer function may be written directly from the standard form as

$$\text{CLTF} = \frac{K}{K + (1 + 0.5s)(1 + s)}$$

$$= \frac{K}{0.5s^2 + 1.5s + (K+1)}$$

Comparison with the standard quadratic factor $s^2 + 2c\omega_n s + \omega_n^2$ and insertion of the selected values of K lead to the following closed loop values:

K	CL poles	ω_n	c
0.1	-1.724	1.483	1.011
	-1.276		
0.5	$-1.5 + j0.866$	1.732	0.866
	$-1.5 - j0.866$		
4.0	$-1.5 + j2.784$	3.162	0.474
	$-1.5 - j2.784$		

As the value of the gain K increases so the system behaviour becomes more oscillatory in nature. This is shown by the increasing imaginary component of the poles, the increasing undamped natural frequency ω_n and the decreasing value of the damping coefficient c. Note that c is related to the stable pole position by $c = \text{Re}/\sqrt{(\text{Re}^2 + \text{Im}^2)}$ where Re and Im are the real and imaginary parts of the poles respectively. Stability is retained at all values of gain.

5.3 Varying gain for system with open loop unstable pole

An open loop system has the transfer function

$$\frac{x(s)}{u(s)} = \frac{K}{(1+0.5s)(s-1)}$$

What is the effect of applying unity negative feedback to this system? Discuss the closed loop pole movement and system behaviour as the gain K is increased, using values for K of 0.5, 1, 4.

Solution This system is open loop unstable with a pole at $+1$. However, over a suitable range of gains, this may be closed loop stable with all poles falling in the left half plane. The closed loop transfer function is

$$\text{CLTF} = \frac{K}{0.5s^2+0.5s+K-1}$$

The poles for the given values of K are as follows:

K	CL poles
0.5	-1.618
	$+0.618$
1.0	-1.000
	0.000
4.0	$-0.500 + j2.398$
	$-0.500 - j2.398$

In this case increasing the gain stabilizes the system. For a range of K the system exhibits a stable overdamped response before oscillatory output occurs at the higher gains. The system is unstable for values of K less than unity.

5.4 Parameter variation and stability

Although pole position is most frequently considered with respect to system gain it is naturally affected by all parameters describing the system's behaviour. In some cases quite small changes may mean the difference between stability and instability. An open loop transfer function is given by

$$\frac{b(s)}{u(s)} = \frac{1}{s^3+2s^2+(4+\delta)s+7}$$

Show that variations in the parameter value $(4+\delta)$ can cause critical changes in the pole positions for both the open and closed loop system.

Solution The *open loop* characteristic equation is

$$s^3 + 2s^2 + (4+\delta)s + 7 = 0$$

Investigation of the roots of this equation, e.g. by the Routh—Hurwitz stability criterion (Chapter 8), shows that it will have unstable roots if $\delta < -0.5$. If δ has this limiting value the roots are at $s = j\sqrt{3.5}$, $s = -j\sqrt{3.5}$ and $s = -2$. Any further negative deviation δ will force the imaginary roots into the right half plane. For less negative values and positive deviations the system is open loop stable with poles in the left half plane.

For the *closed loop* the overall input/output transfer function is

$$\frac{x(s)}{u(s)} = \frac{1}{s^3 + 2s^2 + (4+\delta)s + 8}$$

Now for limiting stability it is necessary that $\delta > 0$ as for all $\delta < 0$ there will be roots of the characteristic equation (closed loop poles) in the right half plane giving instability. With $\delta = 0$ the closed loop poles are $s = j2$, $s = -j2$ and $s = -2$. In this particular case there is a less restricted range on the parameter $(4+\delta)$ to give stability for the open loop system than when the loop is closed.

Thus, absolute stability of a system is not in itself the only consideration that has to be made; it may also be required to assess how this stability is affected by gain and parameter changes. It is necessary therefore to have a suitable 'stability margin' as a measure for control systems (Chapter 9).

System zeros

The zeros of the single-input single-output system are the roots of the equation

$$p(s) = 0$$

They influence the values of the coefficients in the transfer function expansion but do not directly affect the absolute stability of that system which is dependent solely on the poles. However, the *open loop* zeros do affect the position of the *closed loop* poles. In particular, zeros which come in the right half plane (positive real parts) lead to distinctive behaviour and restrict the range of gains for which the closed loop system is stable.

5.5 Effect of position of open loop zero

A system behaviour is described by the input/output relationship

$$6\,\frac{d^2x}{dt^2} + 5\,\frac{dx}{dt} + x = u(t) + a\,\frac{du(t)}{dt}$$

What are the poles and zero(s) of this system? Show how the response of the system varies depending on the magnitude of the coefficient a when the system is subject to a unit step input, i.e. $u(t) = \mathcal{H}(t)$.

Solution Transforming the equation to produce a transfer function gives

$$\frac{x(s)}{u(s)} = \frac{1+as}{(2s+1)(3s+1)}$$

having a zero at $-1/a$ and poles at $-1/2$ and $-1/3$.

For a unit step input $u(s) = 1/s$ so that

$$x(s) = \frac{1}{6} \frac{1+as}{s(s+0.5)(s+0.333)}$$

$$= \frac{1}{s} + \frac{2(1-0.5a)}{s+0.5} - \frac{3(1-0.333a)}{s+0.333}$$

which gives on inversion

$$x(t) = 1 + 2(1-0.5a)e^{-0.5t} - 3(1-0.333a)e^{-0.333t}$$

The effect of the value of a, i.e. of the zero placement, is illustrated by Fig. 5.2.

Note that the presence of the positive zeros, $s = 1$ when $a = -1$ and $s = 0.5$ when $a = -2$, leads to the response setting off in the 'wrong' direction.

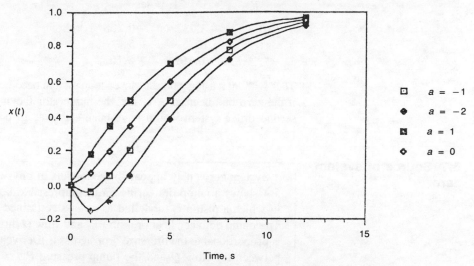

Figure 5.2 Step response variation as system zero position is changed

5.6 Effect of open loop zeros on closed loop behaviour

For the system shown by the block diagram in Fig. 5.3 examine the open loop and the closed loop stability as the gain K is varied using values of K of 0.5, 1.0, 1.5. Contrast this behaviour with that of the system if the open loop zero had instead been negative.

Solution The *open loop* system has poles at -2 and -1 and is thus stable. The positive zero does not affect the stability but leads to the response as demonstrated in the above example for this type of system.

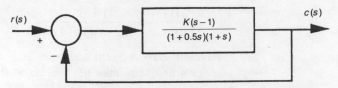

Figure 5.3 System to demonstrate effect of positive open loop zero on closed loop pole position and stability

For the *closed loop* system the transfer function is now

$$\text{CLTF} = \frac{K(s-1)}{K(s-1)+(0.5s+1)(s+1)}$$

$$= \frac{K(s-1)}{0.5s^2+(1.5+K)s+(1-K)}$$

yielding the following poles:

K	CL poles	
0.5	−3.732	−0.268
1	0	−5
1.5	+0.162	−6.162

Thus with increasing gain the closed loop system becomes unstable. In contrast, if the zero had been at -1, i.e. the numerator factor had been $(s+1)$, this second order system would have been stable at all positive values of gain.

5.7 Source of system zero

System zeros may appear in the models of physical systems. Figure 5.4 shows a pump delivering an incompressible liquid into a cylinder in which a piston is modelled as a mass restrained by a linear spring. The pump delivery is at pressure P_1 and flow Q through the inlet valve is proportional to the pressure drop across it. Derive the transfer function between the flow Q and the pump pressure P_1.

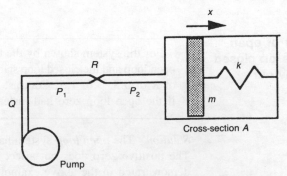

Figure 5.4 Schematic of pump and cylinder

Solution When modelling physical systems it is usually necessary to make simplifying assumptions to produce a readily solved linear relationship, e.g. such as the absence of piston friction and fluid compressibility here. Also the model will only be valid while the suggested Hooke's law model holds for the piston movement.

For flow through the valve restriction

$$Q = \frac{P_1 - P_2}{R}$$

and the velocity of the piston is given by

$$\frac{dx}{dt} = \frac{Q}{A}$$

Force on the piston $= P_2 A = mx'' + kx$.

As the relationship between the pressure P_1 and the flow Q is required eliminate the intermediate variables x and P_2 to yield

$$Q = \frac{1}{R} \left(P_1 - \frac{mQ'}{A^2} - \frac{k}{A^2} \int Q dt \right)$$

giving the transfer function

$$\frac{Q(s)}{P_1(s)} = \frac{A^2}{m} \left[\frac{s}{s^2 + (RA^2/m)s + k/m} \right]$$

This system equation has two poles and a zero at the origin. If the mass and resistance in the cylinder are neglected this reduces to $Q(s)/P(s) = 1/R$. This checks with the system equations, the flowrate becoming proportional to pump pressure.

Problems

1 Plot in the complex plane the position of the poles of the following system models:

(i) $\quad \dfrac{d^2x}{dt^2} + 5\dfrac{dx}{dt} + 6 = u(t)$

(ii) $\quad \dfrac{d^3x}{dt^3} + 5\dfrac{d^2x}{dt^2} = 6\dfrac{dx}{dt} = u(t)$

(iii) $\quad 0.15\dfrac{d^2y}{dt^2} + 0.5\dfrac{dy}{dt} + y = u(t)$

Form the corresponding transfer function and evaluate the response to a unit impulse $u(t) = \delta(t)$ in each case.

Answer (i) $e^{-2t} - e^{-3t}$, (ii) $0.1667 - 0.3333e^{-3t} - 0.5e^{-2t}$, (iii) $3.380e^{-1.667t} \sin 1.97t$

2 Plot in the complex plane the position of the poles of the following system models and comment on the system stability:

(i) $0.125 \dfrac{d^2c}{dt^2} - 0.25 \dfrac{dc}{dt} - c = r(t)$

(ii) $\dfrac{d^3c}{dt^3} + 4 \dfrac{d^2c}{dt^2} + \dfrac{dc}{dt} - 6c = r(t)$

(iii) $\dfrac{d^2x}{dt^2} + \dfrac{dx}{dt} = u(t)$

(iv) $\dfrac{d^2x}{dt^2} = u(t)$

Evaluate the unit impulse response in each case.

Answer   (i) $1.3333(e^{4t} - e^{-2t})$, unstable,
(ii) $-0.3333e^{-2t} + 0.25e^{-3t} + 0.0833e^{t}$, unstable,
(iii) $1 - e^{-t}$, stable,   (iv) t, unstable

3   Figure 5.5. shows a block diagram for a feedback control system. Investigate the position of the open and closed loop poles and hence the open and closed loop stability for the following combinations of controller and plant transfer functions:

(i) $k(s) = 1$ $g(s) = \dfrac{1}{s(s+1)}$

(ii) $k(s) = K\left(1 + \dfrac{1}{10s}\right)$, $K = 1, 2, 3$ $g(s) = \dfrac{1}{(s+2)(s+1)}$

(iii) $k(s) = 3, 4, 5$ $g(s) = \dfrac{1}{(s+4)(s-1)}$

Answer   (i) Stable,   (ii) stable for all K,   (iii) stable for $K > 4$.

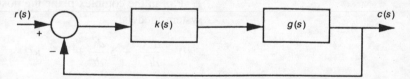

Figure 5.5 Feedback control block diagram

4   The introduction of zeros does not increase the order of the open or closed loop system (provided that it remains proper, i.e. of higher denominator order than numerator order) but it does significantly affect the system dynamics. (This is shown more completely later by the root locus plots.) Compare the step response of the two system equations below. They may be seen as representing the reaction of a second order plant when subjected in open loop (i.e. no feedback of x) to (i) a proportional control input and (ii) an input from a proportional plus derivative action controller.

(i) $\dfrac{d^2x}{dt^2} + 5\dfrac{dx}{dt} + 6 = u(t)$

(ii) $\dfrac{d^2x}{dt^2} + 5\dfrac{dx}{dt} + 6 = u(t) + 0.25\dfrac{du}{dt}$

Answer (i) $0.1667 + 0.3333e^{-3t} - 0.5e^{-2t}$,
(ii) $0.1667 + 0.0833e^{-3t} - 0.25e^{-2t}$

5 Open loop zeros are significant in closed loop system stability. A left half plane zero can be added to improve system response and stability but a right half plane zero can have a much more significant effect and create an unstable feedback system, even at low gains. For the block diagram configuration of Fig. 5.5 again investigate the open and closed loop stability of the following systems:

(i) $k(s) = K(1 + 0.3s)$, $K = 1, 2, 3$ $\qquad g(s) = \dfrac{1}{(s+2)(s+1)}$

(ii) $k(s) = K(1 + s)$, $K = 1, 3$ $\qquad g(s) = \dfrac{1}{(s+2)(s-1)}$

(iii) $k(s) = 3, 6$ $\qquad g(s) = \dfrac{2-s}{(s+1)(s+4)}$

Comment on the relevance of these results in the design and application of controllers for improving the performance of complex systems.

Answer (i) Stable for all K, (ii) unstable for $K = 1$, stable for $K = 3$, (iii) unstable for $K = 3$, $K = 6$.

6 In phase compensators both poles and zeros are deliberately added to the system dynamics to improve system performance. Using the arrangement of Fig. 5.5 again, investigate the effect of the relative positions of the open loop pole and zero of $k(s)$ by evaluating the closed loop poles in the following cases, i.e. different phase compensators with the same 'plant':

(i) $k(s) = K\dfrac{1+3s}{1+0.6s}$, $K = 5$ $\quad g(s) = \dfrac{1}{(s+2)(s+1)}$

(ii) $k(s) = K\dfrac{1+0.1s}{1+0.6s}$, $K = 5$ $\quad g(s) = \dfrac{1}{(s+2)(s+1)}$

(iii) $k(s) = K\dfrac{1+0.6s}{1+10s}$, $K = 5$ $\quad g(s) = \dfrac{1}{(s+2)(s+1)}$

Answer (i) -0.38, $-2.11 \pm j5.09$, (ii) -3.37, $-0.65 \pm j1.74$, (iii) -1.95, $-0.57 \pm j0.17$

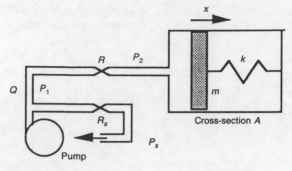

Figure 5.6 Pump and cylinder with relief valve

7 In the system of Fig. 5.6, a development of Fig. 5.4, an attempt is made to smooth out pressure fluctuations at the cylinder valve by fitting a relief valve to return liquid to the pump sump as shown. By considering deviations from an equilibrium condition at which the flow into the cylinder is zero and all the pump flow returns to the sump, derive the transfer function between changes in the pump delivery rate and the change in flow through the relief valve, showing the presence of new system poles.

Answer $\dfrac{ms^2+A^2Rs+k}{ms^2+A^2(R+R_s)s+k}$

8 Although the change in the values of closed loop poles as the open loop gain is changed is looked at more methodically later, it is, nevertheless, instructive to evaluate the plot of such a change for one specific system.

The open loop transfer function for a third order system with unity negative feedback is

$$G(s) = \frac{K}{(1+s)(1+0.5s)(1+0.4s)}$$

(i) Form the closed loop transfer function and the closed loop characteristic equation.

(ii) As K is varied the characteristic equation can be solved to give three roots, real or complex, with their values depending on K. For positive values of K plot the values of these roots (the closed loop poles), placing all three plots on the same complex plane. Show that as K is increased two of the roots become complex and with further increase in K these roots obtain positive real parts.

(iii) Determine the approximate value of K at which two of the roots pass into the real positive part of the plane and explain the closed loop system response in terms of this plot.

Answer (iii) $K = 9.45$

6

Frequency response

Frequency transfer function

So far the response of a system, modelled by a differential equation, to inputs of the impulse, step and ramp form has been evaluated. Of particular interest because of its use in graphical methods and stability analysis is the response of the system to sinusoidal inputs. It has also been a practical alternative for the experimental study of system behaviour.

The frequency response of a system is defined to be the output of that system when it is forced by a steady sinusoidal input of constant amplitude and frequency, and when other initial transients have died down. Under these conditions the output of the linear system is also sinusoidal, possibly expressed as a sum of sinusoids, of the same frequency but with the amplitude and phase with respect to the input governed by the system parameters, including any variable gain. As the frequency varies so do the relative amplitudes and phases. The variation of the open loop amplitude and phase over a range of frequencies gives the frequency response characteristics which may be used for determining closed loop stability (Chapter 9).

For the general single-input single-output linear system described by the transfer function $G(s)$, the response $x(t)$ to a sinusoidal input $u(t) = U \sin(\omega t)$ is expressed by the frequency transfer function

$$G(j\omega) = \frac{X(j\omega)}{U(j\omega)}$$

in which $X(j\omega)$ and $U(j\omega)$ are the Laplace transformed variables $X(s)$ and $U(s)$ with $s = j\omega$.

The ratio of the output amplitude to the input amplitude is given by the magnitude

$$\frac{|X(j\omega)|}{|U(j\omega)|} = |G(j\omega)|$$

and the phase angle between the output and the input is the argument

$$\arg [G(j\omega)] \text{ or } \angle G(j\omega)$$

where a positive angle shows the output to be 'leading' the input. More normally the system will exhibit a phase 'lag' with the argument of $G(j\omega)$, phase angle, negative.

Although of importance in closed loop stability analysis, frequency response is also important in the more general assessment of system performance. For example, it indicates the ability of the system to reduce the effects of high or low frequency noise which may be additional to the desired input/output relationship. At a given frequency the frequency response of a system model may be evaluated in a number of ways, e.g. by the use of phasors to establish the input/output relative gains and phases. There is then no direct solving of equations (although of course full solution of the equations could be used), and simple gain/phase rules are used. For a more general description over a range of frequencies, graphical methods using a polar plot to show the amplitude/phase relationship as frequency is varied may be used, or amplitude and phase plotted more directly against frequency. It is these latter methods which are of particular use in control studies.

Phasor diagrams

For the standard second order dynamic system with oscillatory input of frequency ω and amplitude U the system equation may be written as

$$x + T_1 \frac{dx}{dt} + T_2 \frac{d^2x}{dt^2} = U \cos(\omega t) \quad [\text{or } U \sin(\omega t)]$$

and the corresponding phasor diagram is given by Fig. 6.1.

The phase angle ϕ, by which the output *lags* the input, and the amplitude ratio are given by

$$\phi = \tan^{-1} \frac{T_1 \omega}{1 - T_2 \omega^2}$$

$$\frac{|X|}{|U|} = \frac{1}{\sqrt{[(1 - \omega^2 T_2^2)^2 + \omega^2 T_1^2]}}$$

The frequency ω is required to be expressed in *radians* per second.

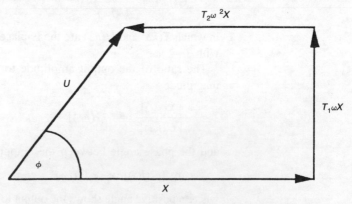

Figure 6.1 Phasor diagram for linear second order system

6.1 Steady state sinusoidal response by calculation and phasor diagram

For the stable systems represented by the following equations establish the steady sinusoidal response at large times, i.e. after initial transients have decayed, for $u(t) = 5\sin(3t)$. Use the magnitude and phase relationships directly and for (i) and (ii) the phasor diagrams also.

(i) $\dfrac{dx}{dt} + 3x = u(t)$

(ii) $\dfrac{d^2x}{dt^2} + 3\dfrac{dx}{dt} + 2x = u(t)$

(iii) $\dfrac{d^3x}{dt^3} + 5\dfrac{d^2x}{dt^2} + 8.25\dfrac{dx}{dt} + 4.25x = u(t)$

Solution (i) Transforming this equation gives

$$sx(s) + 3x(s) = u(s)$$

and

$$g(s) = \frac{x(s)}{u(s)} = \frac{1}{s+3}$$

The gain is

$$|g(j\omega)| = \left|\frac{1}{j\omega+3}\right| = \left|\frac{3-j\omega}{9+\omega^2}\right| = \frac{1}{\sqrt{(9+\omega^2)}}$$

and the phase angle is

$$\angle g(s) = \tan^{-1}\frac{-\omega}{3} = -\tan^{-1}\frac{\omega}{3}$$

For the input $u(t) = 5\sin(3t)$ then $\omega = 3$ and

$$|x(j\omega)| = 5\frac{1}{\sqrt{(9+9)}} = \mathbf{1.179}$$

with a phase angle

$$-\tan^{-1}\tfrac{3}{3} = \mathbf{-45°}$$

i.e. the output lags the input phase by 45°.

The original equation (dividing through by 3) may be represented by the phasor diagram, Fig. 6.2.

(ii) For system equations of higher order the equation should be put into factors. For the second order equation given

$$g(s) = \frac{1}{s^2+3s+2} = \frac{1}{(s+1)(s+2)}$$

The gain $|g(j\omega)|$ is now the *product* of the gains

$$\frac{1}{\sqrt{(\omega^2+1)}} \cdot \frac{1}{\sqrt{(\omega^2+4)}}$$

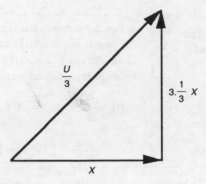

Figure 6.2 Phasor diagram for first order system

and the phase angle is the *sum* of the individual phase angles

$$-\tan^{-1}\frac{\omega}{1} - \tan^{-1}\frac{\omega}{2}$$

For $u(t) = 5\sin(3t)$, $|x(j\omega)| = \mathbf{0.438}$ and the phase angle is $\mathbf{-127.9°}$.

The corresponding phasor diagram, typical of second order systems, is as shown in Fig. 6.3, based on

$$x + 1.5\frac{dx}{dt} + 0.5\frac{d^2x}{dt^2} = 0.5u(t)$$

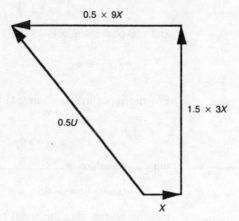

Figure 6.3 Second order diagram

(iii) The transfer function for this third order system factorizes to give a first and a second order factor,

$$g(s) = \frac{1}{(s+1)(s^2+4s+4.25)}$$

to yield the magnitude and phase terms

$$|g(j\omega)| = \frac{1}{\sqrt{(\omega^2+1)}} \cdot \frac{1}{\sqrt{[(4.25-\omega^2)^2+16\omega^2]}}$$

and

$$\angle g(s) = -\tan^{-1}\frac{\omega}{1} - \tan^{-1}\frac{4\omega}{4.25 - \omega^2}$$

giving, for the same input, $|x(j\omega)| = \mathbf{0.123}$ and a phase angle of $-\mathbf{183.2°}$. Note that the phase lag of the second order factor, on account of the sign of its real part when $\omega = 3$, is in the second clockwise quadrant. Hence its value of $-111.6°$ to be added to the first order phase lag of $-71.6°$.

Although the actual figures depend on the system equation the higher the system order the greater the relative attenuation and phase lag at increasing frequencies.

6.2 Frequency response of simple electrical circuits

For the electrical circuits shown in Fig. 6.4 derive the steady state sinusoidal response of the current in the first system and voltage V in the second for unit amplitude inputs.

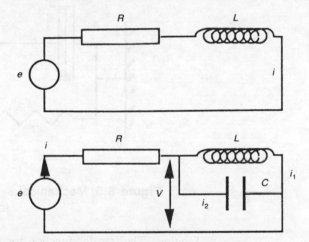

Figure 6.4 Electrical circuits to illustrate frequency response

Solution (i) The transfer function for the resistor−inductor circuit can be shown (Chapter 2) to be

$$g(s) = \frac{i(s)}{e(s)} = \frac{1}{Ls + R}$$

Hence for unit amplitude sinusoidal input the output is given by

$$|i(s)| = \frac{1}{\sqrt{(L^2\omega^2 + R^2)}}$$

with phase angle $-\tan^{-1}(L\omega/R)$, i.e. the current phase lags the voltage phase by this amount.

(ii) For the *RCL* circuit the transfer function is

$$g(s) = \frac{V(s)}{e(s)} = \frac{s}{RCs^2 + s + R/L}$$

Hence the 'output' voltage amplitude $|V(j\omega)|$ is

$$\frac{\omega}{\sqrt{[(R/L - RC\omega^2)^2 + \omega^2]}}$$

with phase angle

$$90° - \tan^{-1}\frac{\omega}{R/L - RC\omega^2}$$

6.3 Mass–spring–damper steady state frequency response

For the mechanical device shown schematically in Fig. 6.5 derive the steady state sinusoidal response given an input $F = 1.\sin(t)$.

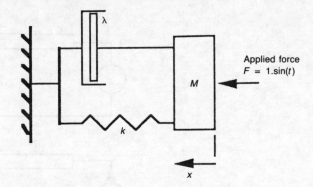

Figure 6.5 Mechanical system with applied sinusoidal force

Solution The transfer function for this system is

$$g(s) = \frac{x(s)}{F(s)} = \frac{1}{Ms^2 + \lambda s + k}$$

so that for the sinusoidal input of unit amplitude and with $\omega = 1$ the output amplitude of steady state oscillations will be

$$\frac{1}{\sqrt{[(k - M)^2 + \lambda^2]}}$$

with phase angle

$$-\tan^{-1}\frac{\lambda}{k - M}$$

At this frequency the amplitude will be at its peak value if $k/M = 1$ and the phase lag will then be $90°$.

6.4 Spring–damper suspension frequency response

A vehicle of mass 1000 kg is modelled as a mass of this size supported on a spring–damper as shown in Fig. 6.6. The spring stiffness is 40 kN m^{-1} and the damping factor of the damper is 80 kN m^{-1} s. Express the frequency response of the system in terms of the outputs, displacement x and the force P, when the surface variation y is the input. What are the amplitude and phase angles of these quantities when y has an amplitude of 0.05 m at an angular frequency of 100 rad s^{-1}?

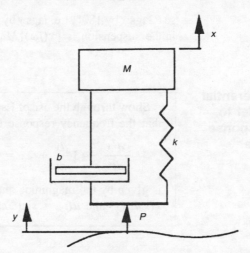

Figure 6.6 Representation of spring–damper suspension

Solution Note that the independent input is the displacement at the base of the spring and damper in parallel. The force transmitted through this 'suspension' is the sum of the forces in these components. Forces and displacements used in the equations are deviations from the steady state equilibrium values. Thus the dynamics are represented by the equation

$$P = k(y-x) + b\left(\frac{dy}{dt} - \frac{dx}{dt}\right) = M\frac{d^2x}{dt^2}$$

Rearranging and expressing as a transfer function leads to

$$g(s) = \frac{x(s)}{y(s)} = \frac{bs + k}{Ms^2 + bs + k}$$

or

$$g(j\omega) = \frac{x(j\omega)}{y(j\omega)} = \frac{jb\omega + k}{-M\omega^2 + jb\omega + k}$$

Hence

$$|g(j\omega)| = \frac{\sqrt{(k^2 + b^2\omega^2)}}{\sqrt{[(k-M\omega^2)^2 + b^2\omega^2]}}$$

with phase angle

$$\tan^{-1}\frac{b\omega}{k} - \tan^{-1}\frac{b\omega}{k - M\omega^2}$$

Substituting in the numerical values $|g(s)| = 0.626$ at a phase angle of **50.9°**. For a surface displacement of amplitude 0.05 m the corresponding amplitude in x is **0.0313 m**.

Using the relationship between x and P

$$P(j\omega) = -M\omega^2 x(j\omega)$$

so P lags x by 180°, i.e. lags y by **129.1°**. The amplitude of the force developed in the suspension is $|x(j\omega)|M\omega^2$, i.e. **313 kN**.

6.5 From differential equation model to frequency response via the transfer function

Show through the use of Laplace transforms and the transfer function that the frequency response for the system described by the equation

$$\frac{d^2x}{dt^2} + 3\frac{dx}{dt} + 2x = u(t)$$

is given by the magnitude and argument of $G(s = j\omega)$ where $G(s) = x(s)/u(s)$ and $u(t) = \sin(\omega t)$.

Solution This example uses a method for the general derivation of the phase and gain expressions but with a particular system equation. The transfer function is

$$g(s) = \frac{x(s)}{u(s)} = \frac{1}{s^2 + 3s + 2} = \frac{1}{(s+1)(s+2)}$$

and

$$x(s) = g(s)u(s)$$

For $u(t) = \sin(\omega t)$ the transformed input is $u(s) = \omega/(s^2 + \omega^2)$ and by using partial fractions

$$x(s) = g(s)\frac{\omega}{s^2 + \omega^2} = g(s)\frac{\omega}{(s+j\omega)(s-j\omega)}$$

$$= \frac{A}{s+1} + \frac{B}{s+2} + \frac{C}{s+j\omega} + \frac{D}{s-j\omega}$$

Inversion leads to

$$x(t) = Ae^{-t} + Be^{-2t} + Ce^{-j\omega t} + De^{j\omega t}$$

As the system is stable the first two terms decrease to zero with increasing time leaving only the two terms with complex exponentials. It is therefore only necessary to evaluate C and D. Returning to the full partial fraction expansion in $x(s)$ and multiplying through by the factor $s+j\omega$ and then putting $s = -j\omega$ leads to

$$C = \frac{g(-j\omega)}{-2j}$$

and using the factor $s-j\omega$ similarly

$$D = \frac{g(j\omega)}{2j}$$

Thus at the 'steady state'

$$x(t) = \frac{g(j\omega)e^{j\omega t} - g(-j\omega)e^{-j\omega t}}{2j}$$

Reduction of this complex expression using $g(j\omega)$ in terms of its magnitude and argument and De Moivre's theorem for the complex exponentials leads after suitable algebra to

$$x(t) = |g(j\omega)|(e^{j(\omega t+\phi)} - e^{-j(\omega t+\phi)})2j$$
$$= |g(j\omega)|\sin(\omega t+\phi)$$

where

$$\phi = \tan^{-1}\left[\frac{\mathrm{Im}\ g(j\omega)}{\mathrm{Re}\ g(j\omega)}\right]$$

That is, the magnitude of the sinusoidal output is $|g(j\omega)|$ times the input magnitude and the phase difference between the input and output is ϕ.

Polar plots

The polar plot is a plot of the magnitude of $G(j\omega)$ and its phase angle as the frequency is taken over its full range from zero to infinity. It may be plotted either from the magnitude and phase evaluated directly or by plotting the real and imaginary parts of $G(j\omega)$ as ω varies, with the polar plot values of magnitude and phase angle being subsequently taken from this cartesian plot. A positive phase angle indicates a phase advance through the system and a negative value a phase lag. Figures 6.7 and 6.8 illustrate basic polar plot characteristics.

In the context of control such a plot is referred to loosely as the Nyquist plot, although in fact it only represents part of the full Nyquist plot used in the stability analysis of feedback systems.

6.6 Simple polar plots

For the two systems described by equations (i) and (ii) plot the polar plot to represent the frequency response of the system. What is the relationship of the magnitude and phase angle values to the real and imaginary parts for the plot?

$$\text{(i)} \quad \frac{dx}{dt} + 2x = u(t)$$

$$\text{(ii)} \quad \frac{d^2x}{dt^2} + 4\frac{dx}{dt} + 3x = u(t)$$

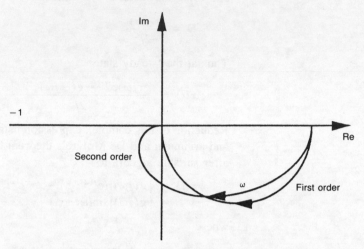

Figure 6.7 Examples of shape of polar plots of first and second order systems (unity steady state gain)

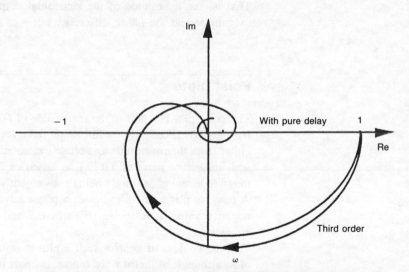

Figure 6.8 Examples of shape of polar plot for third order systems and for systems with a pure delay

Solution (i) The polar plot for this first order system is as shown in Fig. 6.9. Such a plot may be drawn using the magnitude and argument ϕ directly as a true polar plot or the identical information, and shape, shown by plotting using rectangular coordinates where

$$y \text{ coordinate} = \text{Im } g(j\omega) = |g(j\omega)|\sin\phi$$
$$x \text{ coordinate} = \text{Re } g(j\omega) = |g(j\omega)|\cos\phi$$

(ii) This second system equation gives a polar plot over two quadrants showing the maximum phase lag of 180°, Fig. 6.10.

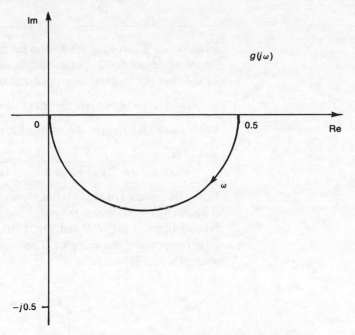

Figure 6.9 Polar plot for first order system

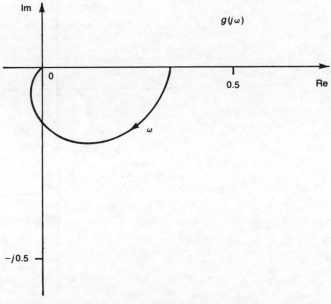

Figure 6.10 Polar plot for second order system

6.7 Third order polar plot

On the basis of (i) the polar plots for the two separate system equations in the previous Example 6.6 and (ii) the magnitude and phase angles directly construct the polar plot for the system equation

$$\frac{d^3x}{dt^3} + 6\frac{d^2x}{dt^2} + 11\frac{dx}{dt} + 6x = u(t)$$

Solution (i) Expressing this system equation in terms of its Laplace transforms (or one could use the D-operator for the derivatives to illustrate the same point) we see that for this input/output equation

$$(s^3 + 6s^2 + 11s + 6)x(s) = u(s)$$

which factorizes to give the transfer function $g(s)$ as

$$\frac{x(s)}{u(s)} = \frac{1}{(s+2)(s^2+4s+3)} = \frac{1}{(s+2)(s+3)(s+1)}$$

This is the product of the two functions in Example 6.6 so that the magnitude at a particular frequency is the product of the individual magnitudes taken from the two figures, Fig. 6.9 and Fig. 6.10, and the phase angle is the sum of the corresponding phase angles. These measurements can then be used for the polar plot in Fig. 6.11.

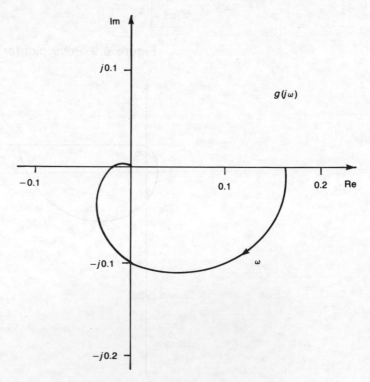

Figure 6.11 Polar plot for third order system

(ii) Alternatively it is seen that the magnitude and phase are given by

$$|g(j\omega)| = \frac{1}{\sqrt{(4+\omega^2)}\sqrt{(9+\omega^2)}\sqrt{(1+\omega^2)}}$$

$$\arg(g(j\omega)) = -\tan^{-1} 0.5\omega - \tan^{-1} 0.333\ \omega$$

and may be evaluated directly, giving the range for this third order system's phase angle from $0°$ to $-270°$.

6.8 Polar plot for system with pure delay

A simple tank type mixer is in series with a length of piping such that an input/output relationship is depicted by the given block diagram, Fig. 6.12.
(i) Draw the polar plot showing the frequency response of this system.
(ii) If the flowrate is decreased, so that the time constant of the mixer becomes 0.625 s and the pure delay in the pipe is also increased to 1.25 s while the steady state gain of the system remains at unity, show the effect of this on the plot and on the response of the system.

$u(s)$ $\dfrac{1}{1+0.5s}$ e^{-s} $x(s)$

Figure 6.12 Representation of simple mixer and pipe flow

Solution (i) For this system of two process elements the combined transfer function is

$$g(s) = \frac{e^{-s}}{1+0.5s}$$

The magnitude and argument for frequency response are respectively

$$1 \cdot \frac{1}{\sqrt{(1+0.25\omega^2)}} \quad \text{and} \quad -\omega - \tan^{-1}0.5\omega$$

The phase contribution ω is in radians. The polar plot is as shown in Fig. 6.13.
As a result of the change in the system parameter (flowrate) the transfer function becomes

$$g(s) = \frac{e^{-1.25s}}{1+0.625s}$$

The increased time constant and the pure delay both lead to an increase in the phase lag produced in this system. The gain and argument terms are now respectively

$$1 \cdot \frac{1}{\sqrt{(1+0.3906\omega^2)}} \quad \text{and} \quad -1.25\omega - \tan^{-1}0.625\omega$$

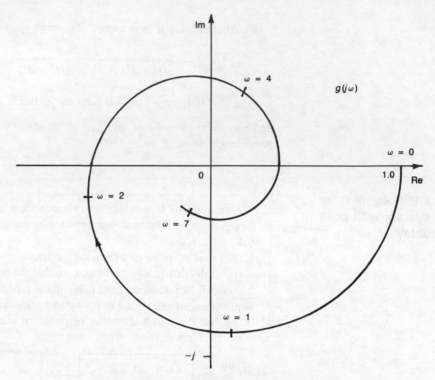

Figure 6.13 Polar plot with pure delay

The new polar plot illustrates the changes in the system frequency response, Fig. 6.14.

In both cases note that the pure 'delay', in this case a 'distance/velocity' delay, has a phase angle without limit as the frequency increases, compared with the simple 'lags' where the phase angles approach a definite limit of multiples of 90°.

6.9 Compensator polar plots

The performance of a control system may be enhanced by the addition of 'phase compensators' which have both poles and zeros. Show the polar plot for the frequency response of the following compensator, noting in particular the quadrant in which it falls and the contributing phase angle:

$$\frac{1+s}{1+0.5s}$$

Combine this with the 'plant' transfer function

$$\frac{2}{s(1+2s)}$$

Further combinations of compensators with system transfer functions are demonstrated in Chapter 10.

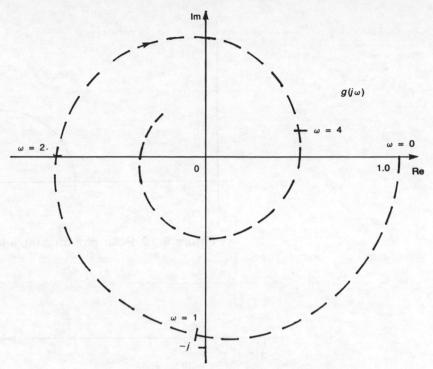

Figure 6.14 Effect of parameter changes

Solution This compensator is a 'phase lead' compensator, the phase angle being positive at all values of frequency. For the polar plot, with $s = j\omega$, the gain and phase are the combination of the values arising from the pole and zero of the function separately. Thus

$$|g(j\omega)| = |1+j\omega| \cdot \left| \frac{1}{1+0.5j\omega} \right|$$

$$= \sqrt{(1+\omega^2)} \cdot \frac{1}{\sqrt{(1+0.25\omega^2)}}$$

with corresponding phase angle

$$\arg(g(j\omega)) = \tan^{-1}\omega - \tan^{-1}0.5\omega$$

The polar plot is as shown in Fig. 6.15.

When this compensator is combined with the plant the effect on the polar plot is to rotate it by a positive angle, reducing phase lag, and to change the magnitude in the frequency range also. The result is shown in Fig. 6.16.

Inverse polar plots

The plot of the inverse of $G(j\omega)$, i.e. of $G^{-1}(j\omega)$, is also used. It is useful in the treatment of systems with certain feedback loop configurations in

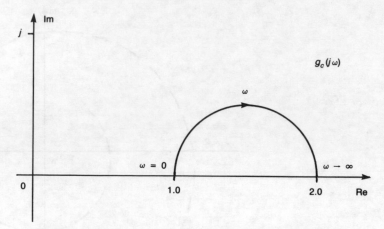

Figure 6.15 Polar plot for simple lead compensator

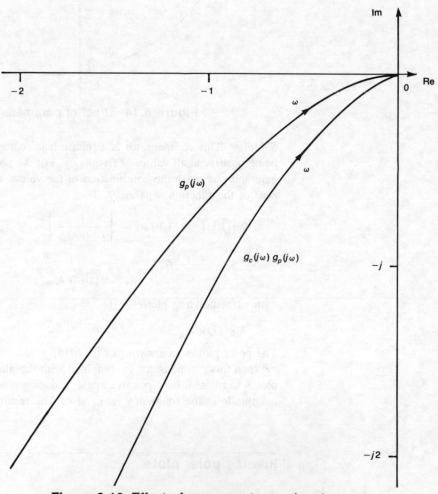

Figure 6.16 Effect of compensator on the plant polar plot

particular and again in stability analysis, with extensions in the field of multi-variable control systems.

6.10 Low order inverse plots

For the two systems described by equations (i) and (ii) plot the inverse polar plot to represent their frequency response:

(i) $\dfrac{dx}{dt} + 2x = u(t)$

(ii) $\dfrac{d^2x}{dt^2} + 4\dfrac{dx}{dt} + 3x = u(t)$

Note the equivalence of this to the direct polar plots given for the same system equations above, Example 6.6.

Solution For (i)
$$g(s) = \frac{1}{2+s}$$

i.e.
$$g^{-1}(s) = 2+s$$

The gain and phase of the function $g^{-1}(j\omega)$ are respectively

$$\sqrt{(4+\omega^2)} \quad \text{and} \quad \tan^{-1}\frac{\omega}{2}$$

Note from Fig. 6.17 that the shape of this plot is quite different from the direct polar plot for the same transfer function, Example 6.6.

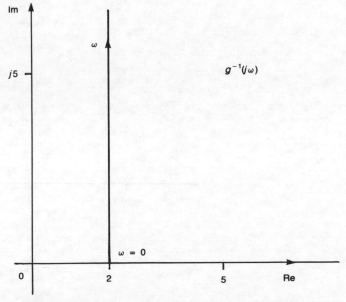

Figure 6.17 Inverse polar plot of a simple pole

For the second order system, (ii), the transfer function is

$$g(s) = \frac{1}{s^2+4s+3}$$

and

$$g^{-1}(s) = (s+1)(s+3)$$

giving modulus and argument now of $g^{-1}(j\omega)$ as

$$\sqrt{(\omega^2+1)} \cdot \sqrt{(\omega^2+9)} \quad \text{and} \quad \tan^{-1}\omega + \tan^{-1}\frac{\omega}{3}$$

shown in Fig. 6.18. The total phase angle is evaluated by considering each factor, first order as here or second order where factorization to first order terms is not possible (underdamped systems).

The phase angles for the direct and inverse plots are of equal magnitude at a given frequency but opposite in sign, hence the plots falling in the quadrants 'reflected' about the real axis. Compare again with Example 6.6.

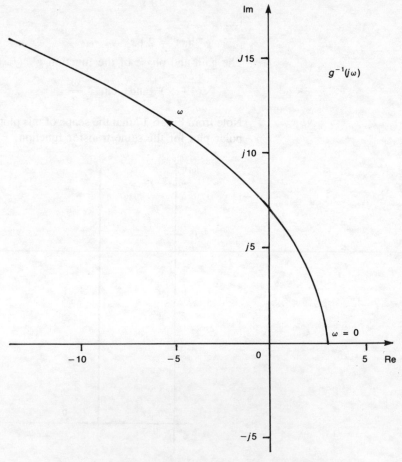

Figure 6.18 Second order inverse polar plot

6.11 Inverse plot with compensator

Plot the inverse polar plots for the first order compensator

$$\frac{1+s}{1+0.5s}$$

Combine this with the 'plant' transfer function

$$\frac{2}{s(1+2s)}$$

and compare with Example 6.9.

Solution The compensator polar plot alone is shown in Example 6.9 together with its magnitude and argument expressions. The inverse plot is shown in Fig. 6.19.

The plant inverse transfer function $g^{-1}(j\omega)$ has gain (magnitude) and phase (argument) respectively

$$0.5\omega\sqrt{(1+4\omega^2)} \quad \text{and} \quad 90°+\tan^{-1}2\omega$$

The transfer function for the two elements in series is

$$G(s) = g_p(s)g_c(s)$$

The overall gain and argument terms for the inverse plot are thus

$$0.5\omega\sqrt{(1+4\omega^2)}\,\frac{\sqrt{(1+0.25\omega^2)}}{\sqrt{(1+\omega^2)}}$$

and

$$90° + \tan^{-1}2\omega + \tan^{-1}0.5\omega - \tan^{-1}\omega$$

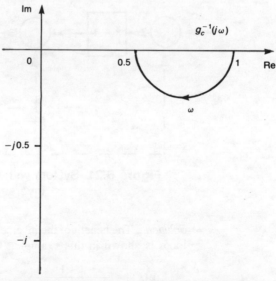

Figure 6.19 Inverse plot for compensator

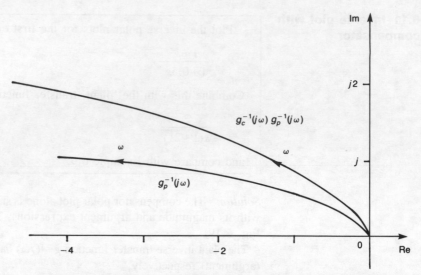

$$g_c^{-1}(j\omega)\, g_p^{-1}(j\omega)$$

$$g_p^{-1}(j\omega)$$

Figure 6.20 Effect of compensator on inverse polar plot
for the plant

Figure 6.20 shows the 'uncompensated' system and the corresponding inverse
plot when the compensator is added in series.

6.12 Inverse plot construction with added inner loop

The block diagram in Fig. 6.21 shows a feedback system. Plot the
inverse polar plot for the frequency response of the *open loop* system.
Using the inverse polar plot investigate the frequency response of this
open loop system if it has a velocity feedback term added as shown.

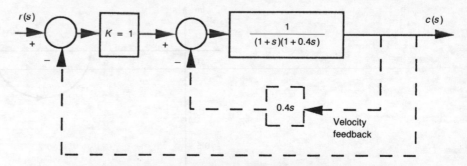

Figure 6.21 System with inner loop suitable for use of
inverse polar plot

Solution The benefit of the inverse polar plot when a system has inner feedback
loops is shown in this example. The 'plant' transfer function is

$$g(s) = \frac{1}{(1+s)(1+0.4s)}$$

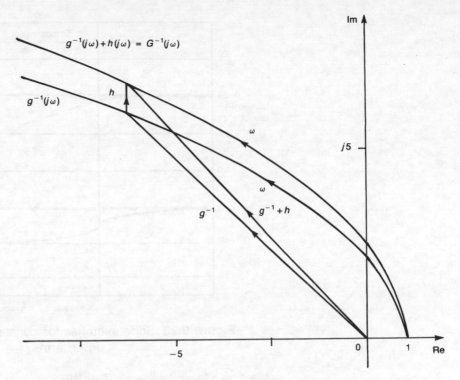

Figure 6.22 Polar plot with the addition of an inner loop

giving

$$|g^{-1}(j\omega)| = \sqrt{(1+\omega^2)} \cdot \sqrt{(1+0.16\omega^2)}$$

and

$$\arg[g^{-1}(j\omega)] = \tan^{-1}\omega + \tan^{-1}0.4\omega$$

If the inner loop with velocity feedback is added, then for this combination

$$G(s) = \frac{g(s)}{1+0.4sg(s)}$$

to give in turn

$$G^{-1}(j\omega) = g^{-1}(j\omega) + 0.4j\omega$$

Thus the addition of the inner loop is handled graphically by a phasor addition. The polar plot for the initial and supplemented system is shown in Fig. 6.22, with the relationship between them shown where $h(j\omega) = 0.4j\omega$.

Logarithmic plots

The logarithmic Bode plot or corner plot is comprised of two graphs drawn to the same independent variable — frequency. The magnitude of $G(j\omega)$ is

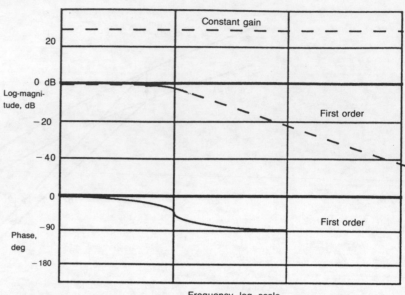

Figure 6.23 Bode sketches for constant gain and first order terms

plotted on a logarithmic scale (base 10) and commonly expressed in decibels and the phase angle is plotted on a linear scale. The independent variable ω is also on a logarithmic scale. For a long time this method of plotting has enabled transfer functions which are of high order but which can be readily expressed as simple factors to be plotted quickly to a good degree of accuracy. With the wider use of computer packages to factorize and plot this type of function the manual plotting requirement has decreased significantly. However, appreciation of the system dynamics and its disposition to instability may still be rapidly obtained by suitable sketching of these plots. Such sketches for Bode plots for simple factors are shown in Figs 6.23 and 6.24.

6.13 Basic Bode plot construction

A dynamic system has the input/output transfer function

$$G(s) = \frac{12}{(s+2)(s+1)(s+3)}$$

Draw the Bode plot and estimate the frequency at which the magnitude of the output is 0.1 that of the steady state (zero frequency) value.

Solution This transfer function is a combination of a constant gain term and three first order lags. The Bode plot, Fig. 6.25, may be accurately sketched by a combination of asymptotes, and the deviation from these at the corner frequencies, to give the log-magnitude plot. Similar approximations may be used for the phase angle or direct, rapid calculation and summation of the phase angles.

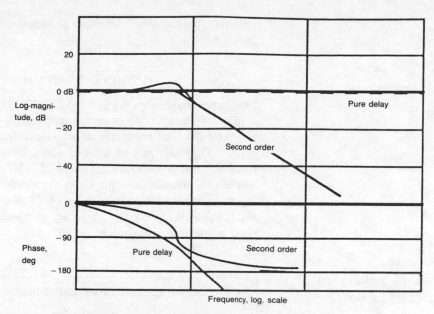

Figure 6.24 Bode sketches for second order terms and a pure delay

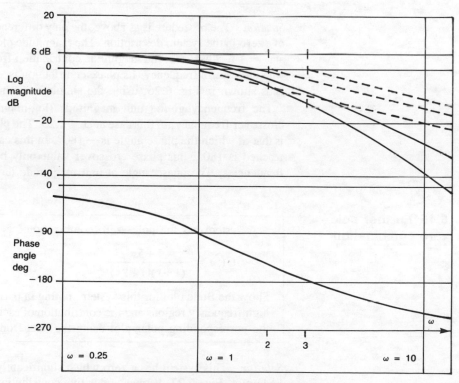

Figure 6.25 Simple Bode plot

Rewrite the transfer function so as to make the numerator the steady state gain, i.e.

$$G(s) = \frac{2}{(1+0.5s)(1+s)(1+0.333s)}$$

The corner frequencies are at 1/0.5, 1, 1/0.333, i.e. at the frequencies equal to (time constant)$^{-1}$. At its corner frequency the log-magnitude of each term is -3.02 dB. The phase angles are respectively $-\tan^{-1} 0.5\omega$, $-\tan^{-1} \omega$, $-\tan^{-1} 0.333\omega$. Each of these is equal to $-45°$ at the respective corner frequency and tends to a final value of $-90°$ at high frequency.

When the magnitude is 0.1 times the steady state value it is 0.2, giving a log-magnitude of 20 log 0.2, i.e. -14. (This is 20 log 0.1 different from the steady state 20 log 2.) The frequency at which this occurs is $\omega = 3.3$. The phase angle is approximately $-180°$.

6.14 Pole–zero combination

A plant plus controller combination has the open loop transfer function

$$G(s) = \frac{2(1+s)}{s(1+0.5s)(1+1.5s)}$$

Draw the Bode plot and determine the frequency at which the gain has unit magnitude. What is the phase angle at this frequency?

Solution The procedure is as above, the only difference now being the addition of a zero in the system description. The magnitude plot has a corner frequency at $\omega = 1$ for this zero, the magnitude contribution from this factor increasing with increase in frequency. Its phase contribution is $+\tan^{-1} \omega$. The Bode plot is as shown in Fig. 6.26, using the simple construction rules.

The frequency giving unit magnitude (log-magnitude = 0) is the gain crossover frequency, in this case at $\omega = 1.25$. The phase crossover frequency is that at which the phase angle is $-180°$. In this case the greatest phase lag reached is 180°, this phase crossover value only being approached at high frequencies. The phase angle at unit magnitude ($\omega = 1.25$) is $-133°$.

6.15 Further pole–zero combination

A system has the pole–zero combination

$$G(s) = \frac{1+5s}{(1+s)(1+2s)}$$

Show the Bode plot for this system, noting in particular the low and the high frequency regions and the contribution of each term to them. Sketch the corresponding polar plot using the data from the Bode plot.

Solution This system has a zero which significantly affects the shape of the Bode plot, Fig. 6.27. It gives a positive contribution to both gain and phase

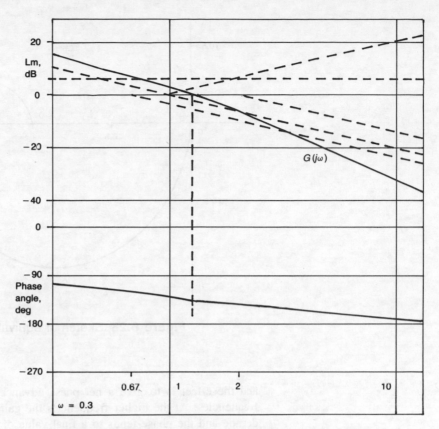

Figure 6.26 Bode plot incorporating system zero

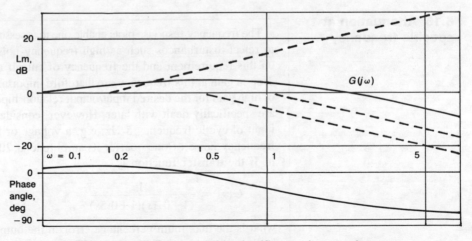

Figure 6.27 System with significant phase advance
characteristic

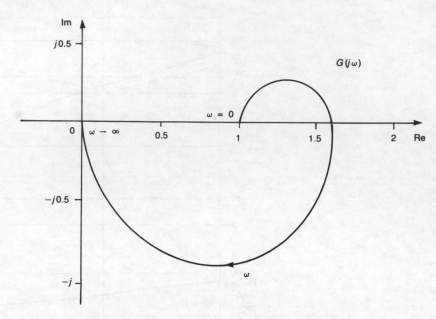

Figure 6.28 Sketch of equivalent polar plot

and the effect is to give a net phase advance in the system at the lower frequencies. At the higher frequencies the gain drops off at just 20 dB per decade and the phase tends to a final value of only $-90°$.

It is possible to sketch quickly the equivalent polar plot taking values directly from the above figure. This shows also the phase advance of the system at low frequencies, the plot falling in the first positive quadrant, Fig. 6.28.

6.16 Attenuation at specific frequency

The frequency response plots enable one to assess the ability of a system to reject disturbances such as high frequency noise. Filters are based on this requirement and the frequency of cut-off and the sharpness of drop in gain at this frequency are therefore important. The requirements of high gain for the desired input/output relationship while rejecting noise are specifically dealt with later. However, consider a system where an input of cyclic frequency 5 Hz, e.g. a voltage or torque, has a higher frequency noise of one-quarter its amplitude at 20 Hz associated with it. If the transfer function is

$$G(s) = \frac{1}{(1+0.1s)(1+0.5s)}$$

what is the maximum percentage error in the output amplitude caused by the unwanted noise? Show how this could be determined from a given Bode plot and from direct calculation from $g(j\omega)$.

Solution For the *dynamic terms* of the system transfer function the log-magnitude and argument are respectively

$$-20\log \sqrt{(1+0.01\omega^2)} - 20\log \sqrt{(1+0.25\omega^2)}$$

and

$$-\tan^{-1} 0.1\omega - \tan^{-1} 0.5\omega$$

The corner frequencies are at $\omega = 10$ and $\omega = 2(\text{rad s}^{-1})$. Note that in this case there is no real need for a Bode plot; the individual gains can be calculated directly at the required frequencies. The *relative* attenuations depend only on the dynamic terms in the transfer functions, not the steady state gain.

The attenuation of the input signal and the noise depends on their respective frequencies. From a Bode plot it could be seen that at 5 Hz ($\omega = 31.4$) the log-magnitude is -34.3 dB, an attenuation of 0.0193. At 20 Hz ($\omega = 125.7$) the log-magnitude is -58.0 dB, an attenuation of 0.001 26. If the noise signal has an amplitude 0.25 times that of the true signal then the ratio of their maximum contributions at the output is

$$\frac{0.25 \times 0.001\ 26 \times 100}{0.0193} = \mathbf{1.62\%}$$

The attenuation figures may also be evaluated directly without recourse to the derived log-magnitude value. The desired ratio is then

$$\frac{0.25\sqrt{[1+(0.1 \times 31.4)^2]} \sqrt{[1+(0.5 \times 31.4)^2]} \times 100}{\sqrt{[1+(0.1 \times 125.7)^2]} \sqrt{[1+(0.5 \times 125.7)^2]}} = 1.62\%$$

6.17 Bode plot and the quadratic factor

Quadratic factors, which factorize to give complex poles and zeros, give a little more difficulty when used with manual methods of Bode plot construction. Although asymptotic values may still be used the area of the plot near to the resonance frequency of the factors needs more attention. For the transfer function

$$G(s) = \frac{2}{(1+0.4s)(s^2+s+1)}$$

draw the Bode plot and estimate the 'gain crossover frequency' and the 'phase crossover frequency'.

Solution The transfer function is again divided into its constituent parts. Substituting $s = j\omega$ the log-magnitude and phase terms are

$$20\log |G(j\omega)| = 20\log 2 - 20\log \sqrt{(1+0.16\omega^2)} \\ - 20\log \sqrt{[(1-\omega^2)^2+\omega^2]}$$

with phase angle

$$-\tan^{-1} 0.4\omega - \tan^{-1} \frac{\omega}{1-\omega^2}$$

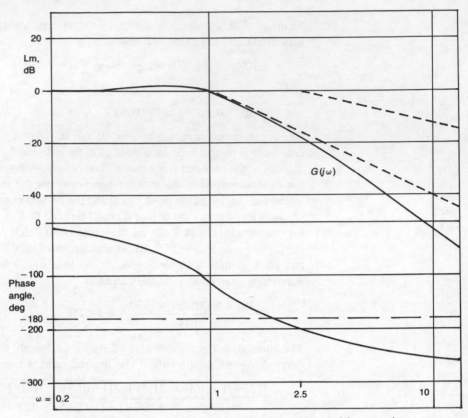

Figure 6.29 Bode plot, third order system with complex poles

The Bode plot follows as in Fig. 6.29.

The gain crossover frequency is at $\omega = 1.00$ and the phase crossover frequency is at $\omega = 1.85$ (rad s^{-1}). Both values are evaluated from the given plots. Direct numerical solution of ω for the attenuation equal to unity and the argument equal to $-180°$ would lead to more precise values.

6.18 Typical computer package plots

The above Bode plots are readily drawn to suitable accuracy using the 'traditional' methods of approximation. They may be readily plotted also by evaluating sufficient points on a hand calculator. In the normal exercise or problem situation this is adequate but with increasing complexity and/or regularity of use, especially in interactive use, such as when varying compensator parameters, computer packages are more suitable. The Bode plots and the polar (Nyquist) plots from such a package are shown for the transfer functions

$$G(s) = \frac{3}{1+s}$$

$$G(s) = \frac{1+0.3s}{(1+0.1s)(1+3s)}$$

$$G(s) = \frac{s^2+s+1}{s(s+1)(s^2+3s+4)}$$

Note that the quadratic factors giving rise to complex poles and zeros cause the user no more difficulty than do the simple poles and zeros.

Solution The solutions to this example are shown directly as reproduced plots from computer printout, Figs 6.30–6.32.

These plots were obtained using Matlab.

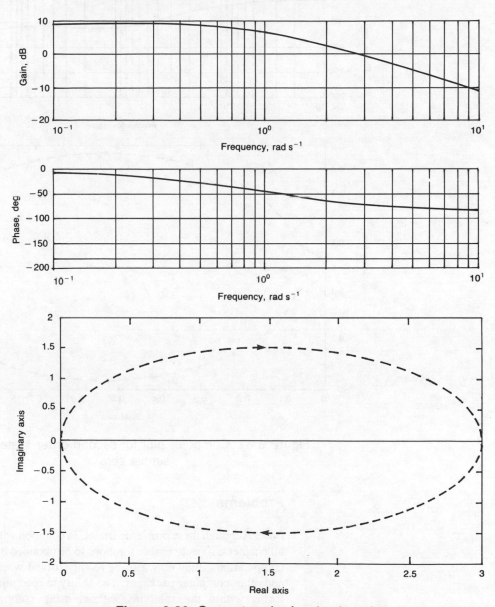

Figure 6.30 Computer plot for simple pole

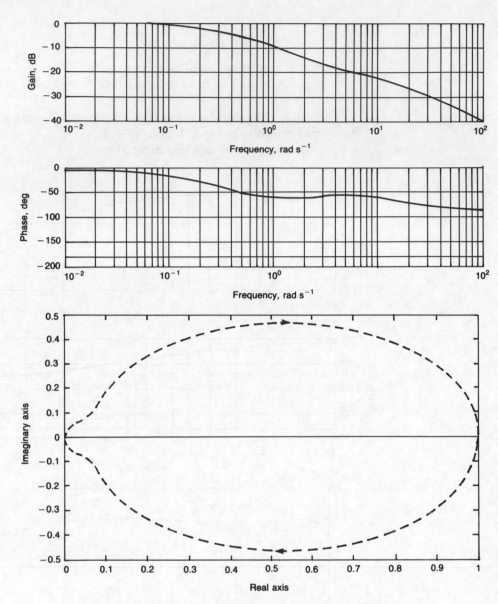

Figure 6.31 Computer plot for second order system with simple zero

Problems

Note. Although these problems are set, in common with others throughout all chapters, so as to enable solutions to be obtained by normal calculator computations, they may also be readily solved when aided by plotting and other computer packages. The ability to recognize and use standard forms within the solutions without using computers is, however, considered to remain important.

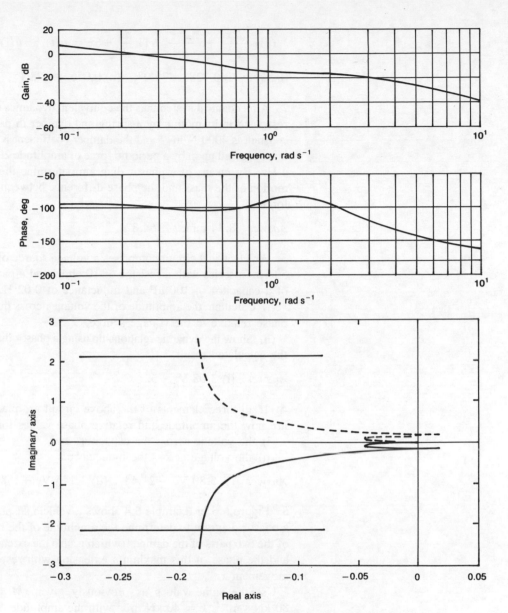

Figure 6.32 Computer plot for fourth order system with complex poles and zeros

1 Use the direct frequency response magnitude and phase relationships to determine the absolute amplitude and the phase angle of x relative to the input u in the following cases:

(i) $\quad 0.2 \dfrac{dx}{dt} + x = u(t),$ $\qquad\qquad\qquad u(t) = 3\sin(20t)$

(ii) $\quad 0.02 \dfrac{d^2x}{dt^2} + 0.2 \dfrac{dx}{dt} + x = u(t),$ $\qquad u(t) = 100\sin(300t)$

(iii) $\dfrac{d^3x}{dt^3} + 6\dfrac{d^2x}{dt^2} + 11\dfrac{dx}{dt} + 6x = u(t), \quad u(t) = 6\sin t$

Answer (i) 0.728, $-76.0°$, (ii) 0.056, $-178.1°$, (iii) 0.6, $-90.0°$

2 A mechanical system has the equivalent dynamics of a mass of 10 kg restricted in its motion by a spring and damper in parallel. The spring constant is 4000 N m^{-1} and the damper coefficient is 25 N s m^{-1}. The mass is acted upon by a periodic force of amplitude 20 N and frequency 4 Hz. By means of a phasor diagram determine the amplitude of the motion of the mass and the phase difference between it and the applied force.

Answer 8.34 mm, $-164.8°$

3 An electrical circuit comprises a voltage source of amplitude 5 V at 50 Hz in series with a resistor of 10 ohms and a parallel combination of a capacitor of 100 μF and inductance of 0.02 H.
 (i) Calculate the amplitude of the voltage across the capacitor and its phase relative to the voltage source.
 (ii) Show the dynamic relationship using a phasor diagram and confirm the calculated result.

Answer (i) 3.08 V, $+52°$

4 If all three elements of the above circuit are placed in series what are now the amplitude and relative phase values for
 (i) the voltage across the capacitor, and
 (ii) the voltage across the inductance?

Answer (i) 5.80 V, $-21.4°$, (ii) 1.145 V, $+158.6°$

5 Figure 6.6 in Example 6.4 shows a vehicle suspension model. Use a phasor diagram to determine the amplitude of the relative movement of the two parts of the damper (which is also the extension of the spring) and the phase of this maximum extension with respect to the vehicle movement x.
 Use the same values as previously given: $M = 1000$ kg, $b = 80$ kN s m^{-1}, $k = 40$ kN m^{-1} with the amplitude of y at 0.05 m and angular frequency 100 rad s^{-1}.

Answer 0.039 m, $+90.3°$

6 Plot or sketch accurately the polar frequency response plots, over the range of all positive frequencies, for the dynamic systems modelled by the following equations or transfer functions:

(i) $\quad 4\dfrac{dx}{dt} + x = u, \qquad\qquad u = \sin(\omega t)$

(ii) $\quad \dfrac{d^2x}{dt^2} + 3\dfrac{dx}{dt} + 2x = u, \qquad u = \sin(\omega t)$

(iii) $g(s) = \dfrac{x(s)}{u(s)} = \dfrac{1}{1+0.3s}$

(iv) $g(s) = \dfrac{x(s)}{u(s)} = \dfrac{1+0.2s}{1+s}$

7 A first order system comprising a simple pole and zero is in series with a pure delay to give the overall transfer function

$$g(s) = \frac{x(s)}{u(s)} = \frac{(1+0.2s)e^{-0.5s}}{1+s}$$

Draw the polar plot for the frequency response of this system, comparing it in particular with problem **6** (iv) above.

8 By considering in particular the high and low frequency regions draw the polar plot $G(j\omega)$ of the systems whose transfer functions are

(i) $G(s) = \dfrac{1}{s(1+s)}$

(ii) $G(s) = \dfrac{1}{s(0.5s^2+s+1)}$

(The real and imaginary parts at low frequency are more informative with regard to establishing the shape of the plot when sketched than the gain and phase alone.)

9 For the systems first introduced in problem **6**, i.e. those with the transfer functions

(i) $g(s) = \dfrac{x(s)}{u(s)} = \dfrac{1}{1+4s}$

(ii) $g(s) = \dfrac{x(s)}{u(s)} = \dfrac{0.5}{1+1.5s+0.5s^2}$

(iii) $g(s) = \dfrac{x(s)}{u(s)} = \dfrac{1}{1+0.3s}$

(iv) $g(s) = \dfrac{x(s)}{u(s)} = \dfrac{1+0.2s}{1+s}$

draw the inverse polar plots. Compare these plots with those of problem **6**, noting how they convey the same magnitude and phase information.

10 Sketch the inverse polar plot $g^{-1}(j\omega)$ for the transfer function

$$g(s) = \frac{(1+0.2s)e^{-0.5s}}{1+s}$$

comparing it with the solution for problem **7**.

11 A second order plant is lightly damped and has the transfer function

$$G(s) = \frac{1}{1+0.1s+s^2}$$

Draw the inverse polar plot for the plant.

Control is initially exercised through standard negative feedback with a proportional action controller having a fixed gain of $K = 8$. Describe the closed loop behaviour of the system.

To improve closed loop performance velocity feedback is added as shown in Fig. 6.21. The velocity feedback gain is set at $k_v = 4$. Use the inverse polar plot to show how this additional control term can be added to produce a new open loop plot for the plant combined with its velocity feedback.

Compare the closed loop performance under the two control arrangements, commenting on the role of the velocity feedback.

12 Although programs are available for the accurate plotting of Bode plots, direct plotting on linear−log graph paper is a progressive exercise leading to normally quite adequate results, especially for low order systems. Draw the Bode plots for the following transfer functions, utilizing the asymptotic magnitude rules and using the results from the earlier sections (i) and (ii) in (iii) and (iv).

(i) $g_1(s) = \dfrac{1}{1+0.4s}$

(ii) $g_2(s) = \dfrac{1}{1+5s}$

(iii) $g_3(s) = \dfrac{5}{s(1+0.4s)}$

(iv) $g_4(s) = \dfrac{5}{s(1+0.4s)(1+5s)}$

Evaluate the magnitude $|g(j\omega)|$ at $\omega = 0.1$ and 5 rad s^{-1} in each case and confirm by calculation. What is the corresponding phase angle at each of these frequencies?

What relationship between the slopes of the log-magnitude plots and the phase angles is indicated by these graphs?

Answer (i) 0.999, $-2.3°$; 0.447, $-63.4°$,
(ii) 0.894, $-26.6°$; 0.040, $-87.7°$,
(iii) 49.96, $-92.3°$; 0.447, $-153.4°$,
(iv) 44.68, $-118.9°$, 0.017, $-241.1°$

13 Draw the Bode plots for the following transfer functions:

(i) $g_1(s) = \dfrac{1}{1+0.4s+s^2}$

(ii) $g_2(s) = \dfrac{1}{s(1+0.4s+s^2)}$

(iii) $g_3(s) = \dfrac{1}{(1+5s)(1+0.4s+s^2)}$

In each case determine from the plots the log-magnitude, and hence absolute magnitude, when the phase angle is $-180°$.

Answer (i) Tends to $-\infty$ dB, zero magnitude as $-180°$ is approached (ii) 8.0 dB, 2.50 (iii) -7.0 dB, 0.445

14 Show the effect on magnitude and phase angle of adding a transfer function zero by drawing the Bode plots for

(i) $g_1(s) = \dfrac{5}{s(1+0.5s)(1+0.1s)}$

(ii) $g_1(s) = \dfrac{5(1+0.2s)}{s(1+0.5s)(1+0.1s)}$

Compare the values of phase lag when $|g(j\omega)| = 1$ (i.e. 0 dB) in the two cases. Note also the effect of the zero on the net phase angle and on the slope of the log-magnitude plot at the higher frequencies. Show that the phase to log-magnitude slope relationship agrees with that demonstrated by problem **12**.

15 The shaping of the Bode plot due to the addition of extra dynamic terms through a controller is used as a key element in controller design. A plant has the transfer function

$$g(s) = \dfrac{1}{(1+0.4s)(1+s)(1+5s)}$$

(i) Draw the Bode plot. A controller of variable gain K is in series with this. What value of K is required so that there is a phase angle of $-140°$ when $|Kg(j\omega)| = 0$ dB?

(ii) The pure gain K is replaced by a controller $k(s)$,

$$k(s) = \dfrac{K(1+1.25s)}{(1+0.15s)}$$

(a) If K is at the value determined in (i) what is the new phase angle when $|k(j\omega)g(j\omega)| = 0$ dB?

(b) If the phase angle when $|k(j\omega)g(j\omega)| = 0$ dB is to be kept at $-140°$ what is the new value of K?

Answer (i) 6.70,
(ii) $-117°$ at $\omega = 1.4$ rad s^{-1}, 12.8 (22 dB) with phase angle of $-140°$ at $\omega = 2.2$ rad s^{-1}

7

Additional controller terms

In Chapter 4 simple feedback control was considered and the most direct form of controller, that incorporating only proportional gain, was used. In this chapter additional controller terms are introduced and the increased flexibility which these give in determining closed loop system behaviour is demonstrated. In Chapters 8 and 9 closed loop stability in particular will be covered and the use of both the time domain and frequency response, as introduced in Chapter 6, extended. Chapter 10 will then extend the general controller form by the use of pole−zero combinations and phase compensators.

Integral action

In the majority of controlled systems the use of proportional control on its own leads to a steady state error between the desired and actual outputs. This may be removed by the addition of 'integral action' which generates a control signal which changes at a rate proportional to the error, and hence continues to integrate up with time. In the presence of a steady demand a zero error is maintained through the application of this action which reaches a constant (non-zero) value. It is not an effective control term in the presence of higher frequency or rapid transient demands and, because it introduces a phase lag, it has a measure of destabilizing effect.

The controller transfer function for proportional plus integral action has the form

$$G_c(s) = k_p + \frac{k_i}{s}$$

However, integral action may be used alone as the sole control action. It is also known as 'reset action' and may be defined in terms of the 'integral action time', k_p/k_i.

7.1 Effect of integral action on error

The effect of adding integral action to remove steady state error may be seen through the application of the final value theorem. A second order system has the open loop transfer function $g(s)$. A controller, $k(s)$, is added in series so that the closed loop transfer function is

$$G(s) = \frac{k(s)g(s)}{1+k(s)g(s)}$$

If $g(s) = 1/(s^2+s+1)$ and $k(s)$ is $k_p + k_i/s$, what is the steady state error of the closed loop system in response to a unit step input when $k_p = 5$ and (i) $k_i = 0$ and (ii) $k_i = 0.2$? What effect will this integral action have on the transient response of the closed loop system?

Solution Consider first the steady state errors using the final value theorem.

(i) With proportional action only, k_p, and a step input the closed loop response $x(t)$ is given its steady state value by

$$x(t)_{s.s.} = \lim_{s \to 0} sx(s) = \lim_{s \to 0} s \frac{5}{s^2+s+6} \frac{1}{s} = \frac{5}{6}$$

Thus the final error between unit input and the output is 1/6.

(ii) With the addition of integral action we have

$$x(s) = \frac{5s+0.2}{[s(s^2+s+1)+5s+0.2]} \frac{1}{s}$$

to give

$$x(t) = \lim_{s \to 0} sx(s) = \frac{0.2}{0.2} = 1$$

In this case the steady state error is zero.

The effect on the transient response is indicated by the closed loop poles without the need for full solution of the transient equation. With integral action the closed loop transfer function is

$$G(s) = \frac{5s+0.2}{[s(s^2+s+1)+5s+0.2]}$$

giving poles at $s = -0.033\,52$, $-0.4832 + j2.3944$ and $-0.4832 - j2.3944$. Without integral action the two poles are at $-0.5 + j2.4367$ and $-0.5 - j2.4367$. Full partial fractions show that the real pole has a small coefficient compared with the real part of the complex pairs. Thus the major oscillatory nature of the response is little changed but there is a slow removal of the error as the output settles to the full input value.

7.2 Integral action with ramp input

What are the equivalent output errors of the system in the previous example if the input is a ramp input $r(t) = 0.1t$?

Solution Although integral action removes the steady state error for the step input this is not so in the case of a ramp, or higher order, input. Proportional action is also unable to cope fully with this input.

(i) As both input and output grow unbounded with time it is necessary to consider the error term directly, i.e.

$$e(s) = r(s) - x(s) = r(s) - k(s)g(s)e(s)$$

to give, for proportional action only,

$$e(s) = \frac{r(s)}{1+k(s)g(s)} = \frac{(s^2+s+1)}{(s^2+s+1)+5} \frac{0.1}{s^2}$$

with the steady state value subsequently, by the final value theorem, tending to infinity. The output value $x(t)$ gets progressively further behind the input $r(t)$.

(ii) With integral action $k(s) = 5+0.2/s$ and

$$e(s) = \frac{(s^2+s+1)s}{(s^2+s+1)s+5s+0.2} \frac{0.1}{s^2}$$

to give similarly, by use of the final value theorem, a steady state error of 0.5. This means that although the input $r(t)$ continues to rise the output almost matches it, reaching a constant finite error of 0.5.

7.3 Use of integral action alone

Show that for the same system as above a stable response may be obtained by the use of integral action on its own. Use $k_i = 0.2$. What is the effect as k_i is increased, with and without proportional action, on the transient behaviour of the system if the proportional gain is fixed?

Solution For the above 'plant' but with integral action only the closed loop transfer function is

$$G(s) = \frac{0.2/s}{(s^2+s+1)+0.2/s} = \frac{0.2}{s^3+s^2+s+0.2}$$

with poles at $s = -0.246$, $-0.377 + j0.790$ and $-0.377 - j0.377$. All these poles are in the left half plane giving a stable system.

As integral action is increased the system will eventually become unstable. Consideration of the roots of the characteristic equation using the Routh criterion shows that as k_i becomes unity in this system so the limit of instability is reached. This might be demonstrated by the use of the root locus diagram. Before this stage is reached the system behaviour will become more oscillatory. This tendency to instability with increasing amounts of integral action is also exhibited when $k_p \neq 0$.

Derivative action

The additional term of derivative action cannot be used on its own and is used in conjunction with proportional action. It gives a control action proportional to the rate of change of error and hence the transfer function for the proportional plus derivative action controller is

$$G_c(s) = k_p + k_d s$$

Its action on a changing error is to reduce overshooting by increasing system damping and thus more rapidly bring a response to within chosen limits near to a desired steady state value following a change in demand. In the presence of high rates of change, however, such as noise, it may cause major fluctuations in control output. It may be expressed in terms of the 'derivative action time', k_d/k_p.

7.4 Effect of derivative action on closed loop poles

Using the same second order system of the earlier examples, illustrate the effect on the closed loop response which is brought about by the addition of derivative action in place of integral action. Use a derivative action constant k_d of 0.5 and $k_p = 5$.

Solution The plant and controller transfer functions are now respectively

$$g(s) = \frac{1}{s^2+s+1} \quad \text{and} \quad k(s) = 5+0.5s$$

The resulting closed loop transfer function is

$$G(s) = \frac{5+0.5s}{(s^2+s+1)+5+0.5s} = \frac{5+0.5s}{s^2+1.5s+6}$$

Using a unit step input as before, $u(s) = 1/s$ and

$$x(t)_{s.s.} = \lim_{s \to 0} sx(s) = \lim_{s \to 0} s \frac{5+0.5s}{s^2+1.5s+6} \frac{1}{s} = \frac{5}{6}$$

This demonstrates that derivative action, because it relies on rate of change, has no effect on steady state error. Use of derivative action alone is thus not a practical proposition.

The two poles of the closed loop system are $s = -0.75 + j2.332$ and $-0.75 - j2.332$. Stability is maintained over any range of k_d, the roots of the characteristic equation always having negative real parts. Without derivative action the closed loop has a damping coefficient of 0.204 and with the introduction of the derivative action this is increased to 0.306, increasing the degree of stability of the system. As in the above examples during a ramp input there is no 'steady state' value but the error continues to grow as with a purely proportional controller.

7.5 Derivative action and noise

Derivative action, because it operates on the rate of change of error, has maximum effect during periods of greatest change within the system. For the system given by the block diagram, Fig. 7.1, show that although the output responds smoothly to ramped up changes in the set point (demand value) the addition of derivative action may reduce the ability of the system to counteract the additional 'noise' input to which a system is subjected.

Use $k(s) = 1+0.1s$ and $g(s) = 1/(s^2+s+1)$.

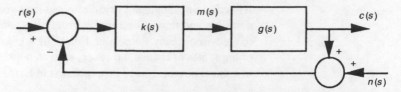

Figure 7.1 Measurement noise in the feedback loop

Solution The higher frequency components, e.g. from measurement noise, are amplified by the call for derivative action. This is shown by looking at the controller output. The closed loop transfer function between the noise and the input to the plant, m, is given by

$$\frac{m(s)}{n(s)} = \frac{-k(s)}{1+k(s)g(s)}$$

(This may be shown by algebraic or block diagram methods.) With the given controller and plant this in turn gives

$$\frac{m(s)}{n(s)} = -\frac{(s^2+s+1)(1+0.1s)}{s^2+1.1s+2}$$

Substituting $s = -j\omega$ enables the frequency response to be determined in terms of phase and magnitude. Considering simply the high frequency range the magnitude of this expression tends to 0.1ω. Thus we may see without further evaluation that the derivative action can lead to high outputs from the controller leading readily to problems of saturation in control elements such as valve movement. The introduction of filters or the alternative phase lead compensator removes this problem which in practice may also be alleviated by the natural high frequency filtering of real system components. The use of velocity feedback as an alternative to derivative action does not suffer in this way, as shown below.

PID control

When all three control actions are combined in a controller this is referred to as a proportional–integral–derivative action controller, or PID for short. It has the capability of producing a combined effect using all three actions although at times one of the modes may be tuned out if required by adjustment of the relevant gain to zero. It is capable of producing both enhanced dynamic and steady state performance. The defining equation for this control action, m, is

$$m = k_p\, e + k_i \int e\, dt + k_d \cdot \frac{de}{dt}$$

and the general block diagram is shown in Fig. 7.2.

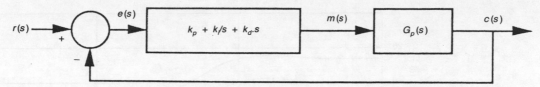

Figure 7.2 General block diagram for PID control

7.6 PI and PD control with second order plant

The benefits of the additional control modes, integral and derivative, are quite different. The derivative term is instrumental more in shaping the dynamic response while the purpose of the integral action is to ensure better steady state and improved slower tracking performance. The effect of these modes is shown in the frequency response plots. Using the basic plant transfer function $g(s) = 1/(s^2+s+1)$ show the changes in the Bode plot resulting from the above integral action and derivative action.

Solution The Bode plots for the plant alone, the effects of proportional control, integral and derivative control are shown in Figs 7.3 and 7.4. At very low frequencies compared with the corner frequencies of the plot the effect of the integral action is most significant showing its steady state influence.

At the higher frequencies the derivative action is dominant.

Note that the effects of integral and derivative action addition are most apparent in distinct frequency ranges on the Bode plot. This makes their combined effect easier to predict. A combination of the two, normally as a

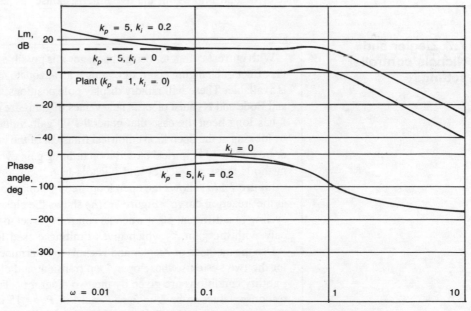

Figure 7.3 Bode plot for second order plant with proportional plus integral action control

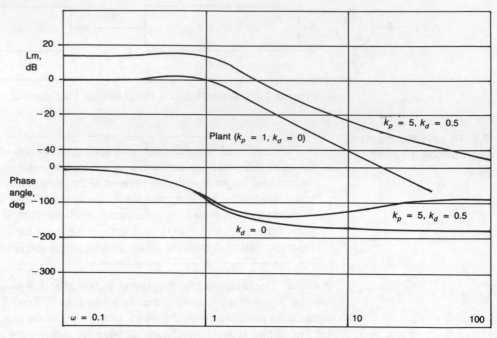

Figure 7.4 Bode plot for second order plant with proportional plus derivative action control

lag—lead controller (Chapter 10), may use both to advantage, removing steady state error and improving the dynamic response.

7.7 Ziegler and Nichols controller settings

With current computer design packages it is possible to see interactively the effect of changing controller gains if an adequate model of the system is available. These will rapidly display pole positions, transient response, and Bode and Nyquist plots as new values of gains are entered. However, it has long been the case that general PID gain values may be derived by looking at the open loop empirical transient of the uncontrolled system or by the use of empirical closed loop limiting gain cycling. Among these methods those of Ziegler and Nichols have been dominant.

Figure 7.5(a) shows the open loop step response for a system, known as the 'reaction curve'. Figure 7.5(b) shows the closed loop continuous cycling of a different plant when its controller is set to proportional mode only with the gain K_u which gives limiting closed loop stability.

Use the methods of Ziegler and Nichols to determine controller settings for the two systems whose open loop transient and closed loop limiting stability conditions are given by the two diagrams. Take corresponding response values to be $L = 3$ s, $T = 14$ s, $P_u = 15$ s with $K_u = 6$ and the output units as °C in (a) and rad s^{-1} in (b). Consider the input m to be in volts in both cases and of unit magnitude.

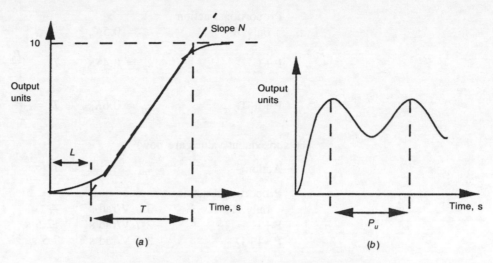

Figure 7.5 Plots for use with Ziegler and Nichols methods

Solution The two main methods of Ziegler and Nichols are (i) the reaction curve method and (ii) the continuous cycling method. The first method uses values from the open loop response of the plant as a basis for the controller settings. In the second method the closed loop response is used. The use of proportional, P+I and P+I+D control is considered in both cases.

(i) For a control input, say m (e.g. volts), and response as in the figure the proposed controller settings are

For proportional (P) action only $\quad k_p = \dfrac{m}{NL}$

P+I $\qquad\qquad\qquad k_p = \dfrac{0.9m}{NL} \qquad T_i = \dfrac{L}{0.3}$

P+I+D $\qquad\qquad\quad k_p = \dfrac{1.2m}{NL} \qquad T_i = \dfrac{L}{0.5} \qquad T_d = 0.5L$

where the integral action time $T_i = k_p/k_i$ and the equivalent derivative action time $T_d = k_d/k_p$. For this system therefore the suggested values would be

Action	k_p	T_i	T_d
Proportional action only, P	0.47 VK^{-1}	—	—
P+I	0.42 VK^{-1}	10 s	—
P+I+D	0.56 VK^{-1}	6 s	1.5 s

(ii) For the continuous cycling (closed loop) method the pure proportional action just to establish undamped closed loop oscillations is determined, K_u. The period then is P_u. Similar to the above the following settings are proposed:

| Proportional action only, P | $k_p = 0.5K_u$ | | |

| P+I | $k_p = 0.45K_u$ | $T_i = \dfrac{P_u}{1.2}$ | |

| P+I+D | $k_p = 0.6K_u$ | $T_i = \dfrac{P_u}{2}$ | $T_d = \dfrac{P_u}{8}$ |

The subsequent values are now

Action	k_p	T_i	T_d
Proportional action only, P	3 V/rad s^{-1}	—	—
P+I	2.7 V/rad s^{-1}	12.5 s	—
P+I+D	3.6 V/rad s^{-1}	7.5 s	1.9 s

Notice that in each case use of integral action requires a reduction in the proportional gain and the addition of derivative action enables the proportional gain to be increased again. This is because of the resulting changes in gain and phase (in terms of the frequency response analysis) and damping which occur with the additional modes. Note that k_p has units dependent on the system and controller.

Velocity feedback

Additional damping may be introduced into a system by the use of 'velocity feedback'. A contribution to the control action is now made which is proportional to the velocity of the output and added to that derived from position feedback. In a block diagram it is represented by an inner loop, Fig. 7.6.

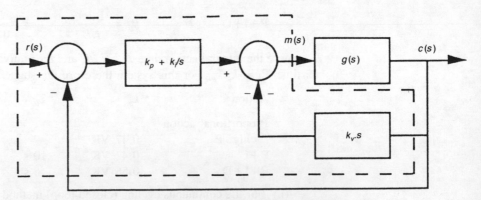

Figure 7.6 Velocity feedback with proportional plus integral action control

7.8 Comparison of derivative action and velocity feedback

Velocity feedback was considered in Chapter 4 where it was seen that it could occur naturally in systems or be introduced by an inner feedback loop. For the undamped second order system given in Fig. 7.7, damping may be introduced by either direct velocity feedback, via k_v, or by derivative action coupled with the proportional controller to give $k(s) = k_p(1 + T_i s)$. If $k_p = 5$, what derivative action time would give the same damping as a k_v value of 0.3? Compare the impulse response for the two closed loop systems and the relative use of each method.

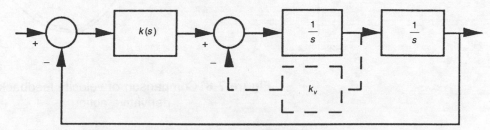

Figure 7.7 Addition of velocity feedback to proportional controller

Solution (i) Establish first the system characteristics with velocity feedback. Reduce the system in stages, i.e. first consider the inner loop which reduces, with $k_v = 0.3$, to

$$\frac{1/s}{1 + 0.3/s} = \frac{1}{s + 0.3}$$

to give the full forward and open loop transfer function, $k(s)/s(s+0.3)$.

With $k(s) = k_p = 5$, the closed loop transfer function is

$$\frac{5}{s(s+0.3)+5} = \frac{5}{s^2+0.3s+5} = \frac{5}{(s+0.15)^2+4.9775}$$

Comparison with the 'standard' second order form in terms of the undamped natural frequency ω_n and the damping coefficient c, $\omega_n^2/(s^2+2c\omega_n+\omega_n^2)$, gives $\omega_n = \sqrt{5}$ and $c = 0.3/2\sqrt{5} = 0.067$.

The response to a unit impulse input is the inverse of the transfer function, i.e.

$$x(t) = e^{-0.15t} \frac{5}{\sqrt{4.9775}} \sin\sqrt{4.9775}t$$

$$= 2.241e^{-0.15t} \sin 2.231t$$

(ii) With derivative action in the forward path series controller the closed loop transfer function is

$$\text{CLTF} = \frac{5(1+T_d s)}{s^2+5(1+T_d s)} = \frac{5(1+T_d s)}{s^2+5T_d s+5}$$

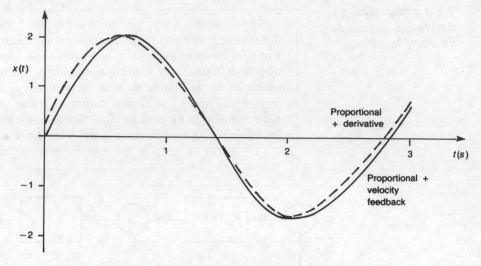

Figure 7.8 Comparison of velocity feedback and
derivative action

Note that ω_n is unchanged at $\sqrt{5}$ but for $c = 0.067$ we require

$$2c\omega_n = 5T_d = 2 \times \sqrt{5} \times 0.067, \text{ i.e. } T_d = \mathbf{0.060}$$

Using this value of T_d to give the unit impulse response, the closed loop transfer function becomes

$$\text{CLTF} = \frac{5(1+0.06s)}{s^2+0.3s+5}$$

$$= \frac{4.955}{(s+0.15)^2+4.9775} + \frac{0.3(s+0.15)}{(s+0.15)^2+4.9775}$$

$$x(t) = \mathrm{e}^{-0.15t}\ (\mathbf{0.3\cos 2.231t + 2.221\sin 2.231t})$$

The damping in each case is the same although there is a difference in the response due to the extra zero in the transfer function, especially at high rates of change, Fig. 7.8.

7.9 Position and speed control of inertia load

Figure 7.9 shows a position control system. Draw a block diagram/ signal flow graph representation and investigate the effect of the addition of velocity feedback from the motor shaft (use a velocity feedback loop gain of 0.0015 V s rad^{-1}) on the transient response. Why is no integral action required?

The same arrangement of motor and load is to be used to drive the load at constant rotational speed using the tachometer feedback only. Draw the block diagram for the proportional control of the load speed. Why should one now consider the introduction of integral action?

The motor has a time constant of 0.5 s and gain of 2.7 N m mA^{-1}, position potentiometers a sensitivity of 9 V rad^{-1}, amplifier gain of 0.6 A V^{-1}, speed reduction ratio n of 16, load moment of inertia of 0.01 N m, load frictional force of 0.02 N m s rad^{-1}. The proportional term of the controller $k(s)$ is 0.01.

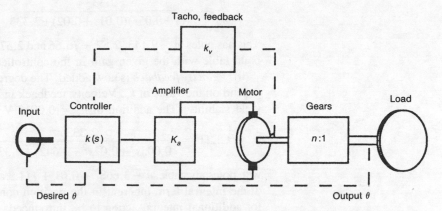

Figure 7.9 Schematic of motor control system

Solution The block diagram is drawn taking into account the physics of the system. In particular note that the motor speed is influenced by the inertia of the load and the distinction that must be drawn between motor speed and load speed. If J is the load moment of inertia, λ the load friction coefficient and T the torque at gearbox output then

$$T = J \frac{d^2\theta}{dt^2} + \lambda \frac{d\theta}{dy}$$

giving

$$\frac{\theta(s)}{T(s)} = \frac{1}{s(0.01s + 0.02)}$$

The transfer function for the motor is $2.7/(1 + 0.5s)$.

The block diagram may now be drawn, Fig. 7.10.

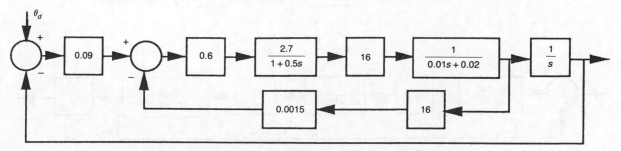

Figure 7.10 Full block diagram for position control

(i) *Without the velocity feedback* the open loop transfer function is

$$\frac{\theta}{\theta_d} = \frac{0.01 \times 9 \times 0.6 \times 2.7 \times 16}{s(1+0.5s)(0.01s+0.02)}$$

giving the closed loop form

$$\frac{\theta}{\theta_d} = \frac{2.333}{s(1+0.05s)(0.01s+0.02)+2.333}$$

This has poles at -9.143, $2.57 + j6.66$ and $2.57 - j6.66$. That is, the system is unstable with the given gain in the controller.

(ii) *Velocity feedback* is now added. The degree of stability introduced will depend on the value of k_v. Velocity feedback in insufficient amounts will not yield stability. The addition of $k_v = 0.0015$ V s rad^{-1} gives

$$\text{CLTF} = \frac{2.333}{0.005s^3+0.02s^2+0.642s+2.333}$$

with poles all stable at -3.668, $-0.61 + j11.3$ and $-0.61 - j11.3$. Because of the integral term inherent in the position control there is no requirement for additional integral action to be introduced.

(ii) With *speed* control and proportional control on the speed error the system is as shown in Fig. 7.11.

As motor speed is fed back but output speed θ is given by the block diagram, the gear box ratio of 16 is used in the block diagram feedback. To obtain correct

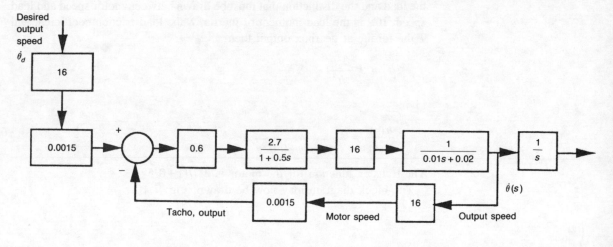

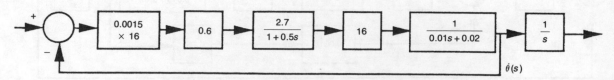

Figure 7.11 Block diagram for speed control

comparison with the motor tacho feedback the input must also be converted to motor speed. The system is now second order in terms of the input/output relationship and

$$\text{CLTF} = \frac{31.104}{(1+0.5s)^2+31.104}$$

there now being an offset between steady state input and output speed values although with the arbitrary given gain the system is very oscillatory. A small amount of integral action will remove the offset and further selection of gain would lead to an improved response.

Feedforward terms

In certain cases it is an advantage to use feedforward terms to improve overall dynamic performance. This amounts to extra generation of control effort based on some knowledge of the system dynamics. This 'feedforward control' uses measured or estimated values of disturbances, interactions and dynamic terms. Because of the uncertainties associated with the dynamics of systems in general the use of such control is insufficient alone and to take account of both these uncertainties and unmeasured disturbances feedback control must be retained with it. As a result feedforward control is somewhat restricted in its applications.

7.10 Disturbance 'removal' with feedforward control

The signal flow graph of Fig. 7.12 represents a process subject to a disturbance entering at a 'midpoint' between the feedback/controlled input and the output.

(i) Show how in theory the effect of this disturbance may be eliminated with feedforward control. What is the transfer function between any residue of this disturbance and the output?

(ii) Show also that conflict can arise when trying to reduce the effect of such disturbances, e.g. at measurements, within a feedback control system.

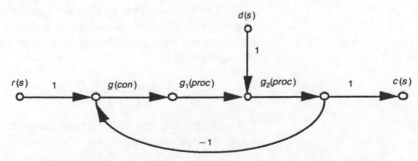

Figure 7.12 Signal flow graph for system with process disturbance

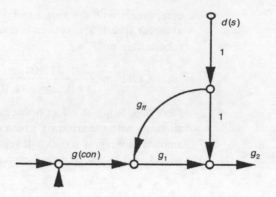

Figure 7.13 Addition of feedforward control

Solution (i) The addition of feedforward control is shown by the addition to the signal flow graph, Fig. 7.13.

The ideal is to add a signal at a control input so that the effect of this at the point of insertion of the disturbance is sufficient to cancel it out. Thus equating the disturbance and its fed forward compensation gives

$$g_1(s)g_{ff}(s)d(s) = -d(s)$$

i.e.

$$g_{ff}(s) = -g_1^{-1}(s)$$

This requires accurate knowledge of the process or mechanism, plus the ability to measure the disturbance, coupled with the ability to generate control terms which will probably contain derivative terms of order greater than unity because of the nature of $g^{-1}(s)$. These factors give difficulty in application and are greatly restrictive to use.

If the full effect of $d(s)$ is not removed by the feedforward addition then a residual effect will be felt at the output. Between $d(s)$ and $c(s)$ in the forward path transfer function is $g_2(s)$ and the closed loop transfer function is

$$\frac{c(s)}{d(s)} = \frac{g_2(s)}{1 + g_c(s)g_1(s)g_2(s)}$$

(ii) The ability to restrict sensitivity to disturbances within different parts of a loop may conflict. Consider a disturbance at the measurement, added at a node in the feedback. The closed loop transfer function between this and the output will be

$$\frac{c(s)}{d_m(s)} = \frac{g_c(s)g_1(s)g_2(s)}{1 + g_c(s)g_1(s)g_2(s)}$$

If the disturbance had been represented at the plant output, i.e. added after $g_2(s)$, the equivalent would be

$$\frac{c(s)}{d_2(s)} = \frac{1}{1 + g_c(s)g_1(s)g_2(s)}$$

At any one frequency one of these cannot be made very small without the magnitude of the other tending to unity.

7.11 Feedforward using controller set point

To cope with unexpected fluctuations in the primary flowrate to a solvent mixing tank the control scheme shown in Fig. 7.14 is proposed. It is an example of feedforward control exercised through coupling two controllers, the set point of the controller C_1 being adjusted by the change in flowrate in the prime solvent flow. Such a correction to the controller C_1 is intended to prevent the build up of imbalance in the tank before the concentration/monitoring controller at the outlet itself applies corrective action. Draw a descriptive block diagram showing the principle of control with and without this addition to the control system. Establish an overall simplified block diagram relating the output concentration to deviations in the primary flow.

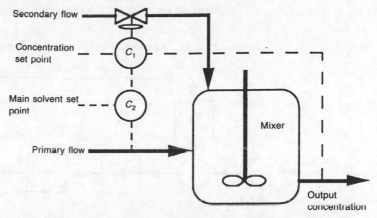

Figure 7.14 Control of mixer with secondary flow set point set by primary flow

Solution The block diagrams for the system with and without the second controller are shown in Figs 7.15 and 7.16 respectively.

If the dynamics of small deviations can be represented as linear and with

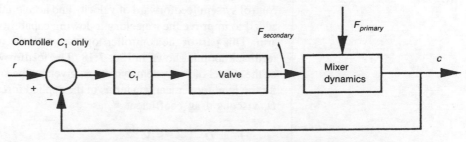

Figure 7.15 System with single controller only

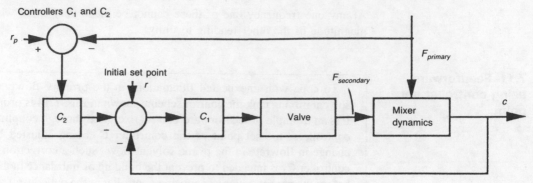

Figure 7.16 System with combined controllers

transfer functions $g_1(s)$ and $g_2(s)$ then a block diagram can be drawn representing the simplified system, e.g. on the lines of Fig. 7.17.

A deviation in primary flow will give an output from C_2 changing in turn the output from C_1 and the secondary flow. To establish overall control, though, an outer control loop is still required.

Controllers C_1 and C_2, equivalent block diagram for small deviations

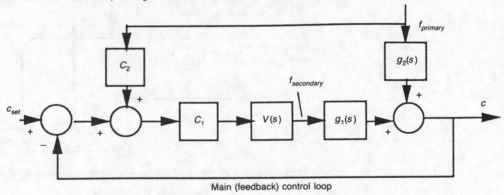

Figure 7.17 Block diagram for combined controllers

7.12 Feedforward control using simple model

Feedforward control has been proposed to improve the dynamic response of systems in other ways. For example, as part of an enhanced control system feedforward of velocity and acceleration demand has been added to improve the trajectory following capability of a link of a robot arm. This part of the controller together with the fundamental position feedback loop is shown in Fig. 7.18. The feedforward terms are based on the model of the system dynamics, given by the simplified relationship for the joint movement q in terms of the applied force or torque F, inertia D, viscous drag coefficient V, as

$$F = D\,\frac{d^2q}{dt^2} + V\frac{dq}{dt}$$

Establish the basic transfer system for the position control loop and show how this is changed by the addition of the feedforward terms if the model is a good representation of the link dynamics.

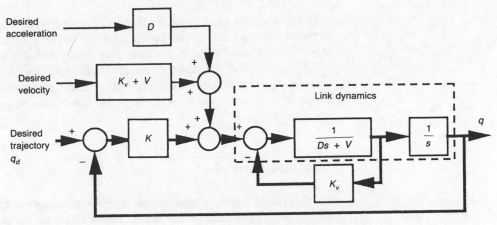

Figure 7.18 Single robot link control

Solution (i) Consider the principal feedback loop with the velocity feedback. The inner loop transfer function is $1/(K_v+Ds+V)$ to give

$$\frac{q(s)}{q_d(s)} = \frac{K}{s(Ds+V+K_v)+K}$$

and the *position error* $e(s) = q_d(s) - q(s)$ is given by

$$\frac{e(s)}{q_d(s)} = \frac{s(Ds+V+K_v)}{s(Ds+V+K_v)+K}$$

The error in velocity is $dq_d/dt - dq/dt$ and is expressed in transforms as $sq_d(s) - sq(s)$. The subsequent algebra using the above transfer function gives the *velocity error* from

$$\frac{sq_d(s)-sq(s)}{q_d(s)} = \frac{s^2(Ds+V+K_v)}{s(Ds+V+K_v)+K}$$

(ii) Adding the velocity feedforward term $q_d'(K_v+V)$ (see block diagram) gives the additional closed loop contribution to $q(s)$ at the output of

$$\frac{s(V+K_v)q_d(s)}{s(Ds+V+K_v)+K}$$

The *position net error* ratio, from both position feedback and the effects of this extra feedforward on the loop, then becomes

$$\frac{e(s)}{q_d(s)} = \frac{s(Ds+V+K_v)+K-s(V+K_v)}{s(Ds+V+K_v)+K}$$

$$= \frac{Ds^2}{s(Ds+V+K_v)+K}$$

With this added term the *velocity error* (transformed and with respect to the desired trajectory $q_d(s)$) reduces to

$$\frac{Ds^3}{s(Ds+V+K_v)+K}$$

(iii) Adding the acceleration feedforward gives a closed loop transfer function between this and the output position of $Ds^2/[s(Ds+V+K_v)+K]$, and the position error (and velocity error) is thus finally reduced to **zero**.

Adding these feedforward terms improves the trajectory following provided there is a good model of the dynamics. In practice even if the model contains errors in D, V, etc., the level of the performance may be increased, even if not ideally.

Cascade control

Overall performance of quite simple single-input single-output systems may be enhanced by the use of two controllers in which the set point of one is set by the output of the other. Such a combination is known as 'cascade control'. Each controller has its own feedback signal and the control is not dependent on a model of the system dynamics in the same way as feedforward control. It is relevant to point out that although this structure originates from the use of distinct physical controllers, with the broad acceptance of digital (computer based) control such functions now form an easily implemented structure in a single controller. The principle of this method is shown in Fig. 7.19.

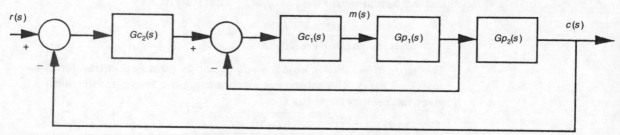

Figure 7.19 Principle of cascade control

7.13 Cascade control using controller set points

Consider the solvent mixing problem again. In an alternative scheme, Fig. 7.20, the primary flow is now controlled by a flow controller and there are two feedback loops. The prime objective is still to produce constant concentration within the vessel and to do this the composition controller adjusts the set point of the flow controller. If the flow controller is proportional action only and the composition controller proportional plus integral action, derive the overall closed loop transfer function for the composition control using a model based on perfect mixing within the vessel. What effect would the removal of the inner loop of the flow controller have on overall system performance?

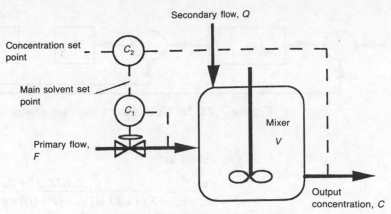

Figure 7.20 Cascaded controllers

Solution The solution is best effected through combining the simple mixer transfer function and the control loops in a block diagram. By block diagram reduction or simple algebra the overall transfer function is derived.

The plant transfer function is between the output concentration and the changes in the primary flow. Although the need for the control may be because of changes in the secondary flow Q of pure additive, assumed small, the control loop is a relationship between the primary flow F and output concentration C. Consider a small change in F of f and a corresponding change in C of c. A mass balance at some initial steady state condition yields the equation

$$Q = C(F+Q)$$

After the disturbance in F of f the additive balance becomes

$$Q = V\frac{dc}{dt} + (C+c)(F+f+Q)$$

so that on subtracting the steady state equation and neglecting the second order term fc,

$$V\frac{dc}{dt} + c(F+Q) = -\frac{Qf}{F+Q}$$

Defining the initial concentration $Q/(F+Q) = C_o$ the transfer function for the mixer is

$$g(s) = \frac{c(s)}{f(s)} = \frac{C_o}{Vs+(F+Q)}$$

The valve may have its own dynamics but if these are fast compared with the controller and mixer then it can be represented by a constant gain L and the block diagram incorporating this is as shown in Fig. 7.21.

Reducing the inner loop to a single function gives an overall open loop transfer function

$$k_p\left(1+\frac{1}{T_is}\right)\frac{KL}{1+KL}\frac{C_o}{Vs+(F+Q)}$$

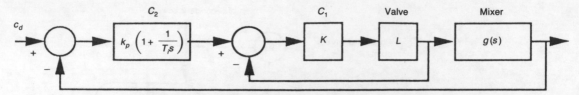

Figure 7.21 Block diagram for cascade controllers

with closed loop

$$\frac{c(s)}{c_d(s)} = \frac{k_p KLC_o(1+T_is)}{T_i(1+KL)s[Vs+(F+Q)]+k_p(1+T_is)KLC_o}$$

If the inner loop is removed so that the two controllers are replaced by just one proportional plus integral controller then the overall closed loop transfer function becomes

$$\frac{c(s)}{c_d(s)} = \frac{k_p(1+T_i)Lg(s)}{T_is+k_p(1+T_i)Lg(s)}$$

$$= \frac{k_p(1+T_i)LC_o}{T_is[Vs+(F+Q)] + LC_ok_p(1+T_is)}$$

In both cases the system is second order but the introduction of the second loop affects the dynamic gain and response rate, reducing the effect of fluctuations in the primary flow.

7.14 Signal flow graph for cascade controllers

For the heated vessel shown in Fig. 7.22 a cascade control system is installed to reduce the effect of the lag between changes in the service stream, the steam to the jacket, and the resulting change in the temperature of the reactor contents. Draw a signal flow graph showing the interaction between the components of a simple representation of this system.

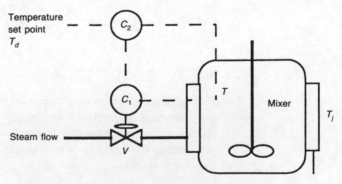

Figure 7.22 Temperature control with cascade controllers

Solution It is possible to show the interrelation of the system components and dynamics without explicitly deriving the individual transfer functions. Thus the signal flow graph structure may be shown as in Fig. 7.23, further extension requiring the usual process of modelling to develop the equations and transfer functions which relate changes in the mixer temperature to set point and steam changes.

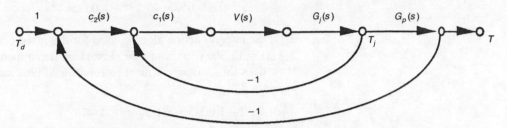

Figure 7.23 Signal flow graph for cascade temperature controller

Problems

1 A proportional controller plus plant, with negative unity feedback, has the open loop transfer function

$$k(s)g(s) = \frac{10}{0.25s^2 + 1.25s + 1}$$

An integral action addition to the controller adds a factor to the numerator of $(1 + 0.2/s)$. What is the closed loop 'steady state' error (i.e. as $t \to \infty$) of the system without and with integral action if the demand input is

(i) a unit impulse,
(ii) a unit step, and
(iii) a unit ramp?

Answer (i) 0, 0,   (ii) 0.0909, 0,   (iii) ∞, 0.5

2 A servo mechanism has its dynamics represented by the second order transfer function

$$\frac{\theta(s)}{T(s)} = \frac{1}{s(5s + 1.5)}$$

where θ is angular displacement (rad) and T is applied torque (N m). It is driven by a control system having a transfer function

$$\frac{T(s)}{e(s)} = 25\left(1 + \frac{1}{2s}\right)$$

with unity negative feedback.

(i) Explain, showing a block diagram, why the integral action is required when the mechanism is used as a speed control system, i.e. with velocity as the feedback variable, but not when it is used as a position control system.

(ii) Evaluate the closed loop unit step response for the system acting as (a) a position control system with proportional action and (b) a speed control system with proportional plus integral action.

Answer (ii) $1 - 1.002e^{-0.15t} \sin(2.231t + 1.5037)$
$1 + 0.0525e^{-0.5234t} - 1.0525e^{-4.777t}$

3 If the integral action alone is used for the speed control in problem **2**, $k(s) = 12.5/s$, what is now the closed loop step response of the system? How does this compare with the proportional plus integral control of problem **2**?

Answer $1 - 1.0045 \sin(1.574t + 1.476)$

4 The same mechanism of problem **2**, i.e.

$$\frac{\theta(s)}{T(s)} = \frac{1}{s(5s + 1.5)}$$

is too lightly damped in the closed loop position control case with the proportional action control. Derivative action is added so that the controller becomes

$$k(s) = 25(1 + 0.8s)$$

(i)   Show that this gives approximately critical closed loop damping.

(ii)   What is the effect of the extra term on the steady state values?

(iii) Evaluate the step response and compare the closed loop pole positions with those for the proportional controller alone.

Answer (iii) $1 - e^{-2.15t} [1.789 \cos(0.6145t + 0.2784) - 2.620 \sin(0.6145t + 0.2784)]$

5 A device is modelled as two first order lags in series with

$$g(s) = \frac{20}{(1 + s)(1 + 6s)}$$

It is to be controlled by a proportional controller $k(s)$ of proportional gain $K = 20$ to give acceptable steady state error.

(i) Plot the frequency response Bode plot of $k(j\omega)g(j\omega)$ to determine the phase angle of $k(j\omega)g(j\omega)$ when $|k(j\omega)g(j\omega)| = 1$, i.e. 0 dB.

(ii) The difference between this phase angle and $-180°$ is the phase margin. It may be improved by adding derivative action in the controller so that, with a derivative action time T_d of 0.3 s, for example,

$$k(s) = K(1 + 0.3s)$$

With K unchanged at 20 what is the new 'phase margin' as determined using the Bode plot?

(iii) What disadvantage may come from this addition?

Answer (i) $-172°$ at $\omega = 8.1$ rad s^{-1}, (ii) $84°$

6 To remove steady state error fully in problem **5** integral action is introduced with an integral action time T_i of 5 s and at the same time the gain K is reduced to 15. Derivative action is not included. The open loop transfer function is now

$$k(s)g(s) = \frac{20 \times 15(1+5s)}{5s(1+s)(1+6s)}$$

Show the result of these changes on the Bode plot and determine the new phase angle (and phase margin) when the magnitude of $|k(j\omega)g(j\omega)|$ is again 0 dB.

Answer $-172°$ ($8°$ phase margin) at $\omega = 7.0$ rad s^{-1}

7 Discuss the merits of using integral and derivative action control modes to improve closed loop performance. Why is it frequently possible to consider the addition of these two modes almost independently?

Combine the derivative action and integral action terms in problems **5** and **6** as a PID controller. Use $K = 15$. Show the effect of this controller by using the Bode plot.

8 A feedback control system has the plant transfer function

$$g(s) = \frac{1}{s(1+0.4s)}$$

and the controller has constant gain, i.e. $k(s) = 10$. Draw a signal flow graph to illustrate the addition of velocity feedback with gain k_v. What is the closed loop transfer function for the full system?

If the velocity feedback gain is $k_v = 2$ compare the unit step response of the system with and without the velocity feedback term.

Answer $1-1.512e^{-3.75t} \sin(3.307t+0.723)$
$1-1.033e^{-1.25t} \sin(4.841t+1.318)$

9 A lightly damped system has the transfer function

$$g_p(s) = \frac{1}{1+6s+100s^2}$$

and a proportional controller K with unity negative feedback.

(i) If $K = 10$ what is the closed loop steady state error for a unit step input and what are the closed loop damping coefficient and the closed loop pole positions?

(ii) Velocity feedback is added (as in Fig. 7.6). Draw the block diagram for this system. What is the transfer function for $g_p(s)$ and the velocity feedback combined if the velocity feedback gain is k_v?

(iii) What value is required for k_v (with K unchanged) if the closed loop damping coefficient is to be 0.6?

(iv) Draw the Bode plots (i) and (iii) to show the effect on the frequency response of this added control term.

Answer (i) 0.0909, 0.0904, $-0.03 \pm j0.330$

(ii) $\dfrac{1}{100s^2 + (6+k_v)s + 1}$ (iii) 33.8

10 Figure 7.12 shows a general representation of a disturbance entering a process. Draw the equivalent block diagram for this system with the addition of feedforward control to cancel the effect of this disturbance. Use the following transfer functions:

$$g_{con}(s) = K = 10 \qquad g_1(s) = \frac{1}{1+s} \qquad g_2(s) = \frac{1}{1+2s}$$

(i) What are the closed loop transfer functions between the disturbance $d(t)$ and the output $c(t)$ and between the additional feedforward contribution $f(t)$ and the output $c(t)$?

(ii) In the steady state what percentage of a constant disturbance d is felt at the output in the absence of the feedforward control?

(iii) The feedforward controller is modelled with an inaccuracy such that

$$g_{ff}(s) = -(1+0.9s)$$

Plot the unit step response $c(t)$ to the disturbance $d(t)$, (a) in the absence of feedback, (b) with feedback present.

Answer (i) 0, $\dfrac{g_1(s)g_2(s)}{1+g_1(s)g_2(s)g_{con}(s)}$ (ii) 9.09%

(iii) $0.1(e^{-0.5t} - e^{-t})$, $0.0225e^{-0.75t} \sin(2.222t)$

11 Figure 7.19 shows the principle of cascade control. Explain the difference between this and feedforward control and the purpose of the inner loop. Within such a system the controllers are respectively proportional plus derivative and integral so that

$$g_{C1}(s) = \frac{1}{T_i s} \qquad g_{C2}(s) = K(1+T_d s)$$

and the contributing plant transfer functions are

$$g_{p1}(s) = \frac{1}{1+5s} \qquad g_{p2}(s) = \frac{1}{1+50s}$$

Derive the overall transfer function between the concentration set point r and the output concentration c. What is the overall steady state gain? Discuss how the individual controller terms might be set.

Answer $K/(1+K)$

12 A mixer is surrounded by a heater jacket as shown in Fig. 7.22.

Why is the cascade control scheme proposed to reduce the effects of uneven steam flow, rather than a single loop from the in-tank temperature sensor to the valve position control?

Draw a block diagram to represent this system's simplified dynamics. Assume that the valve action has a first order lag of time constant T_v, the vessel jacket temperature responds to the steam flow as a first order system with time constant T_s, and that there is a first order lag, time constant T_j, between the jacket temperature and the temperature of the mixer contents. Take all of these lags to have unity steady state gain. Fluctuations in pressure in the steam supply add an input to the system which in turn would add a proportional flow through the (uncontrolled) steam valve. Add this input to the block diagram. By block diagram manipulation determine

(i) the overall transfer function between the temperature set point and the tank temperature, and

(ii) the transfer function form between the steam supply perturbations and the temperature of the tank contents.

Answer (i) $\dfrac{c_1(s)c_2(s)}{(1+T_vs)(1+T_ss)(1+T_js) + c_1(s)(c_2(s) + 1 + T_js)}$

(ii) $\dfrac{1+T_vs}{(1+T_vs)(1+T_ss)(1+T_js) + c_1(s)(c_2(s) + 1 + T_js)}$

8

Stability and the time domain

The stability of a closed loop system may be assessed through use of the 'time domain' or 'frequency domain'. The former is considered in this chapter.

Stability of closed loop systems

The stability, or instability, of a closed loop system is shown by the location of its poles in the complex plane. These are dependent on the system's open loop poles and zeros and on gain. A system which is open loop stable may, depending on its order and structure, become unstable with negative feedback. However, the prime objective of feedback control is to ensure stability, the finer points of performance then being accounted for by adjustment of controller gains and parameters. With linear systems stability is a global property. Once the criterion for stability has been established, e.g. gain, at some operating point or for some input, then this confirms stability for all conditions. For non-linear systems this is not necessarily so.

For a system having constant parameters such that the open loop poles and zeros do not vary, the poles of the closed loop system are located in the complex plane according to the system (controller) gain. Variation of the gain thus determines the stability. This may be shown graphically by plotting the closed loop poles, the roots of the closed loop characteristic equation, as the gain is varied, giving rise to a 'root locus' plot, and determining the gain at which the poles move into the right half (real part positive) of the plane. Alternatively, this condition for limiting gain may be determined non-graphically.

Routh—Hurwitz criterion

With the availability of root finding programs which can be run on low cost computers or hand calculators some of the more 'mechanistic' methods which have found applicability assume less importance. For day-to-day problem solving, these methods are still useful in illustrating the onset of instability at conditions of limiting gain and in finding the limiting gain values. The Routh—Hurwitz stability criterion is based on the relationships between the coefficients of an algebraic equation if all roots of that equation are to have negative real parts. Through the relationship of the roots of the characteristic equation to system stability it is thus possible to determine stability without full solution of the equation.

8.1 Use of the Routh–Hurwitz criterion for stability determination

As was illustrated in Chapter 5 stability of a system, or at least of its model, is determined by the position of its poles. The stability of a closed loop (or open loop) system may be determined by use of the Routh–Hurwitz stability criterion without full solution of the characteristic equation. Investigate the stability of the systems represented by the block diagrams in Figs 8.1 and 8.2 and the following transfer functions:

(i)

(a) $\qquad g(s) = \dfrac{3}{s^2+3} \qquad\qquad h(s) = \dfrac{1}{s+1}$

(b) $\qquad g(s) = \dfrac{1}{s^3+s^2+3s+1} \qquad h(s) = \dfrac{1}{s^2+1}$

(ii) $\qquad g_1(s) = \dfrac{1}{s+1} \qquad\qquad g_2(s) = \dfrac{2}{s+2}$

$\qquad\qquad h_1(s) = 0.5s \qquad\qquad h_2(s) = \dfrac{1}{s+1}$

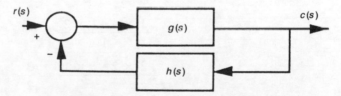

Figure 8.1 Basic negative feedback system

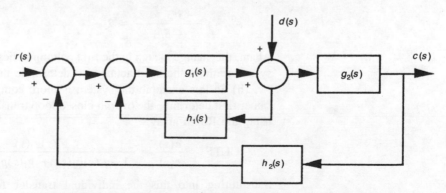

Figure 8.2 Negative feedback with inner dynamic loop

Solution The relationships for the coefficients of a polynomial to be such that its roots all have non-positive real parts may be written out directly in terms of those coefficients. However, a general method of establishing this uses the Routh array as illustrated here.

(i) (a) For the given $g(s)$ and $h(s)$ the CLTF is

$$\text{CLTF} = \frac{3(s+1)}{(s^2+3)(s+1)+3}$$

with characteristic equation $s^3+s^2+3s+6 = 0$. The array is set up from the coefficients and these are manipulated according to the set rules to give

s^3	1	3
s^2	1	6
s^1	$\dfrac{1 \times 3 - 1 \times 6}{1} = -3$	0
s^0	$\dfrac{-3 \times 6 - 1 \times 0}{-3} = 6$	

The change of sign indicates a root with positive real part in the closed loop characteristic equation, i.e. an unstable closed loop system.

(b) All the open loop poles are stable, being in the left half of the complex plane. The closed loop characteristic equation is now

$$s^5 + s^4 + 4s^3 + 2s^2 + 3s + 2 = 0$$

and the array is

s^5	1	4	3
s^4	1	2	2
s^3	$\dfrac{1 \times 4 - 1 \times 2}{1} = 2$	$\dfrac{1 \times 3 - 1 \times 2}{1} = 1$	
s^2	$\dfrac{2 \times 2 - 1 \times 1}{2} = \dfrac{3}{2}$	$\dfrac{2 \times 2 - 1 \times 0}{2} = 2$	
s^1	$\dfrac{3/2 \times 1 - 2 \times 2}{3/2} = \dfrac{-5}{3}$		
s^0	$\dfrac{-5/3 \times 2}{-5/3} = 2$		

Again, the change in sign in the first column indicates an unstable closed loop system without the need actually to determine the system roots.

(ii) In this example the system is more complex but the method is the same; first determine the overall closed loop transfer function and then check using the Routh array.

$$\text{CLTF} = \frac{c(s)}{r(s)} = \frac{g_1(s)g_2(s)}{1+g_1(s)h_1(s) + g_1(s)g_2(s)h_2(s)}$$

Substituting into this the individual transfer functions given yields the characteristic equation $1.5s^3+5.5s^2 + 6s + 4 = 0$ to yield the array

$$\begin{array}{llc}
s^3 & 1.5 & 6 \\
s^2 & 5.5 & 4
\end{array}$$

$$s^1 \quad \frac{5.5 \times 6 - 1.5 \times 4}{5.5} = 4.909$$

$$s^0 \quad 4$$

Thus in this case all closed loop poles are stable, illustrating that apparent complexity may not itself be an indication of system stability or otherwise. Note that the criterion (for closed loops) for stability is based on the roots of the mathematical closed loop characteristic equation directly and is subsequently used whether or not the open loop is stable or unstable. (Nyquist's criterion is also based on the position of the closed loop roots but is based on mappings. It also covers the difficulties which arise for non-minimum phase systems, i.e. those with open loop poles and zeros having positive real parts.)

8.2 Use of the 'auxiliary equation'

To cope with 'non-standard' terms appearing in the Routh array additional steps are made. Determine the stability conditions for the systems having the closed loop characteristic equations

(i) $s^4 + s^3 + s^2 + s + K = 0$

(ii) $2s^3 + 2s^2 + s + 1 = 0$

Solution (i) Apply the Routh array to this system to give

$$\begin{array}{llll}
s^4 & 1 & 1 & K \\
s^3 & 1 & 1 & \\
s^2 & 0 & K &
\end{array}$$

Replace the zero in the first column by a small positive ϵ,

$$\begin{array}{lll}
s^2 & \epsilon & K
\end{array}$$

$$s^1 \quad \frac{\epsilon - K}{\epsilon}$$

$$s^0 \quad K$$

Now as ϵ tends to zero the penultimate term in the table tends to $-K/\epsilon$, i.e. requiring K to remain negative if there is to be no change of sign here. On the other hand the final term requires K to remain positive. Hence the closed loop system is unstable for all K.

(ii) The array in this case is

$$\begin{array}{lll}
s^3 & 2 & 1 \\
s^2 & 2 & 1 \\
s^1 & 0 & 0
\end{array}$$

This row indicates repeated roots given by the auxiliary equation $2s^2 + 1 =$

0, the coefficients coming from the previous line. The roots are at $s = \pm j\sqrt{0.5}$ and on differentiation of the auxiliary equation the row in the array obtains the coefficients

$$
\begin{array}{lll}
s^1 & 4 & 0 \\
s^0 & 1 &
\end{array}
$$

There is no further indication of unstable roots and the system is marginally stable. (Note that for low order cases like this direct use of the relationships between the equation coefficients which come from the basis of the array could be quicker to use.) For the second, third and fourth order systems these are as follows where the general equation is

$$a_n s^n + a^{n-1} s^n + \ldots + a_1 s + a_0 = 0$$

In all cases all coefficients must be non-zero and positive and:

Second order   No further requirement
Third order $a_2 a_1 - a_3 a_0 > 0$
Fourth order $a_3 a_2 - a_4 a_1 > 0$ and $a_1(a_3 a_2 - a_4 a_1) - a_3 a_0 > 0$

Root locus

The root locus plot shows how the roots of the closed loop characteristic equation, i.e. the closed loop poles, vary as a system parameter, usually a controller gain, is varied. It not only indicates absolute stability but also predicts general closed loop performance. The roots are plotted in the complex plane and this may be done by complete evaluation of them for a number of gain settings. Alternatively they may be plotted using a number of 'construction' rules to enable a rapid assessment to be made without the necessity for full root evaluation. The root locus method is used widely, whether drawn 'by hand' using these construction rules or by computer solution.

8.3 Basic root locus diagram

> For the system in Fig. 8.2 the variable gain K is introduced by using the transfer functions
>
> $$g_1(s) = \frac{1}{s+1} \qquad g_2(s) = \frac{2}{s+2}$$
>
> $$h_1(s) = 0.5s \qquad h_2(s) = K$$
>
> *Sketch* the root locus plot to show the effect of changes in the gain K on the placement of the closed loop poles. Outline how such plots may be obtained by 'trial and error' satisfaction of the magnitude and argument conditions.

Solution For the system in question the open loop transfer function is

$$\frac{g_1(s)}{1+g_1(s)h_1(s)} \times g_2(s) \times h_2(s)$$

i.e.

$$\frac{2K}{(1+1.5s)(s+2)(s+1)}$$

The root locus is shown in Fig. 8.3. The open loop poles are at $-1/1.5$, -2, -1. The asymptotes at $60°$, $180°$, $300°$ intersect at -1.22 on the real axis and the breakpoint is at -0.82. The branches cross the imaginary axis at $\pm 2j$ at the critical value of $K = 10$.

The general rules to help sketching without full computation appear in standard texts and these are used explicitly in the next example. However,

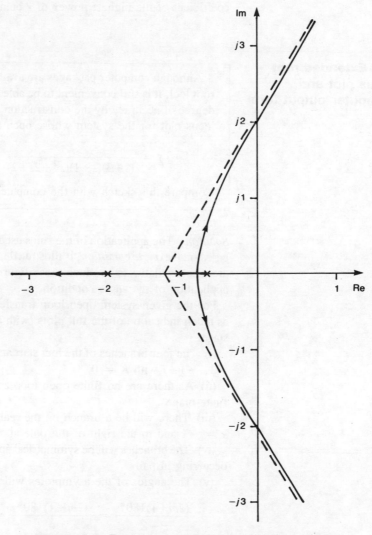

Figure 8.3 Root locus plot for third order system

successive computations may be made until the points making up the plot satisfy the angle condition

$$\Sigma \arg(s - z_i) - \Sigma \arg(s - p_i) = (2k+1)180°$$

The gain along the plot is given by

$$K = \frac{\Pi \| (s - p_i) \|}{\Pi \| (s - z_i) \|}$$

Thus from specific points s_i along the branches the gain K to give these closed loop pole values s_i may be determined. Note that the K refers to the value of gain when the open loop is expressed such that the coefficient of the highest power of s is unity. This is when the factors are expressed as $(s+a)$ etc. and the closed loop characteristic equation is then a 'monic polynomial' with the coefficient of the highest power of s being unity.

8.4 Extended root locus plot and computer output form

Although computer packages are available for plotting the closed loop root loci, it is still convenient to be able to plot such loci to a reasonable degree of accuracy by the construction rules of Evans. Sketch the root locus plot for the system whose open loop transfer function is

$$g(s) = \frac{K}{(s+3)(s+1)(s^2+2s+2)}$$

Compare this sketch with the computer derived plot.

Solution The application of the rules used for sketching the root locus plots, based on the *angle condition*, is illustrated in the following. Computer packages are also available to produce these figures by equation root solving or by application of the angle condition.

For the given system open loop transfer function we proceed to produce as much indication of the full plots (with as little effort as possible!) at each stage.

(i) The four branches of the loci start at the four open loop poles, $s = -3$, -1, $-1 \pm j$ with $K = 0$.

(ii) As there are no finite open loop zeros no branches will finish in the finite plane.

(iii) There will be a branch on the real axis to the left of the real pole at $s = -1$ and to the right of the pole at $s = -3$.

(iv) The branches will be symmetrical about the real axis, any complex poles occurring in pairs.

(v) The angles of the asymptotes will be given by

$$\frac{(2m+1)180°}{P-Z} = \frac{(2m+1)180°}{4} = 45°, 135°, 225°, 315°$$

(vi) The intersection of the asymptotes is given by

$$\sigma = \frac{\Sigma \operatorname{Re}(p_i) - \Sigma \operatorname{Im}(z_i)}{P - Z} = \frac{-3 - 1 - 1 - 1}{4} = -1.5$$

(vii) The intersection of the branches with the imaginary axis may be determined by the angle condition directly or by use of the Routh–Hurwitz criterion to give both critical gain and the crossing point on the axis. This yields, using the methods demonstrated earlier,

$$K_c = 18.9 \text{ and } s = \pm j 1.53$$

(viii) If the open loop poles and zeros are expressed in the general form $s = -a_i \pm jb$ the breakaway point at $-q$ is given by

$$\sum_{\text{zeros}} \frac{q - a_i}{(q - a_i)^2 + b_i^2} - \sum_{\text{poles}} \frac{q - a_i}{(q - a_i)^2 + b_i^2}$$

$$= \frac{1}{q - 1} - \frac{1}{q - 3} - \frac{2(q - 1)}{(q - 1)^2 + 1} = 0$$

and solution of this cubic gives the root on the loci at $q = 2.4$, i.e. the breakaway point at -2.4.

(ix) Applying the angle condition to determine the direction of the branch departure from the upper complex open loop pole gives

$$\phi\ dep = (2m + 1)180° - 90° - 90° - \tan^{-1} 0.5 = -26.6°$$

and $26.6°$ for the lower branch.

Although additional points may be added by the angle condition directly the above is enough to show the root loci accurately, Fig. 8.4.

Use of a computer based package leads to the plot of Fig. 8.5.

8.5 Calculation of gain for specific closed loop damping

Using the root locus method for the system having the closed loop transfer function

$$G(s) = \frac{K}{(s + 15)(s + 45) + K}$$

determine the gain K so that the closed loop system has a damping factor half that of critical damping.

Solution Inspection, coupled with a knowledge of the standard closed loop/open loop relationship, shows that the given closed loop equation arises from the system with open loop transfer function

$$g(s) = \frac{K}{(s + 15)(s + 45)}$$

The root locus diagram has the simple form in Fig. 8.6, at low gains being overdamped and never becoming unstable. To determine the gain at damping

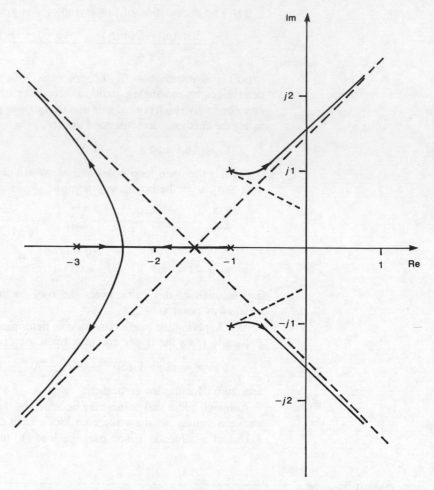

Figure 8.4 Root locus plot for fourth order system using construction rules

factor 0.5 draw the added line at $\cos^{-1} 0.5$ to the negative real axis from the origin, the intersection giving the closed loop poles corresponding to this. The required value of gain is then calculated from the magnitude condition at this point in the complex plane. The intersection is at $-30 + j51.96$. The gain K corresponding to this is AB.BC (see example above),

$$K = \text{AB.BC} = \sqrt{(51.96^2 + 15^2)} \cdot \sqrt{(51.96^2 + 15^2)} = \textbf{2925}$$

This is a simple system example of the general method making use of the magnitude condition. Here it is easily confirmed by calculation.

The characteristic equation is

$$s^2 + 60s + 675 + K = 0$$

and with $c = 0.5$, $\omega_n = 60$. We require that $675 + K = \omega_n^2$, i.e. $K = 2925$ as above.

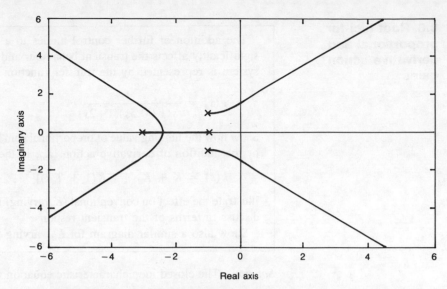

Figure 8.5 Root locus plot for fourth order system using computer program

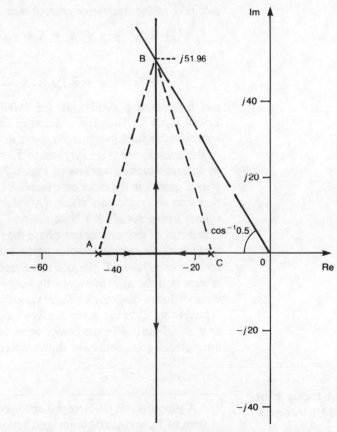

Figure 8.6 Use of root locus plot to determine gain for given damping

8.6 Root loci for proportional and derivative action gains

The addition of further control modes to a proportional controller significantly affects the transient behaviour and stability. A third order system is represented by the transfer function

$$g(s) = \frac{1}{s(1+0.5s)(1+2s)}$$

Show how the limiting value of proportional gain K for stability is affected by the addition of derivative action, K_d, in the controller,

$$k(s) = K + K_d s \; [= K(1 + T_d s)]$$

Illustrate the effect on conventional (K varying) root locus diagrams and discuss in terms of the transient response.

Show also a similar diagram for K_d varying for a constant K value.

Solution The closed loop characteristic equation with proportional action is

$$s^3 + 2.5s^2 + s + K = 0$$

The limiting value of K for stability, e.g. by Routh–Hurwitz, is $K<2.5$. Addition of the derivative control term gives the characteristic equation

$$s^3 + 2.5s^2 + s + K + K_d s = 0$$

or

$$s^3 + 2.5s^2 + s + KT_d s + K = 0$$

and the limiting conditions for stability now are given by $K>0$ and $2.5(1+K_d)>K$. Thus the addition of the derivative action enables K to be increased (on stability grounds) above its previous maximum value. The effect of this added zero, $(1+T_d s)$ where $T_d = K_d/K$, on the system is illustrated by the root locus diagrams in Fig. 8.7, using $T_d = 0$ and $T_d = 0.1$. The critical gain K in the latter case increases to 3.33 before the diagram branches pass into the right half plane. (A value of $T_d = 0.4$ and above leads to a system stable for all K.) Note the new asymptotes.

In terms of the transient response the figure shows that a limit is placed on the fastest pole but the complex pair are moved further into the left hand plane for a given value of K giving a more rapid response. As the derivative action is increased the system eventually becomes stable for all K. For a K of 3.33 the root locus diagram as K_d is varied is shown in Fig. 8.8.

Increasing T_d at the given K gives a critical value for stability of $T_d = 0.1$ as in this figure. For the combination of $K = 3.33$ and $T_d = 0.1$ the same three closed loop roots are shown on each of the two diagrams.

8.7 Root locus plot with added integral action

A proportional plus integral action controller, with a proportional gain term of K_p and variable integral action time T_i, is connected to a plant having unity steady state gain and a simple first order time constant of 60 s. In the feedback loop the instrumentation also acts as a simple lag, but with a time constant of 15 s. The steady state gain in this instrumentation is unity. What is the relationship between K_p and T_i for stability to be maintained? If T_i is set at 6 s sketch the root locus diagram and evaluate the step response of the system when the gain K_p is set, using this plot, to give a damping factor of 0.5.

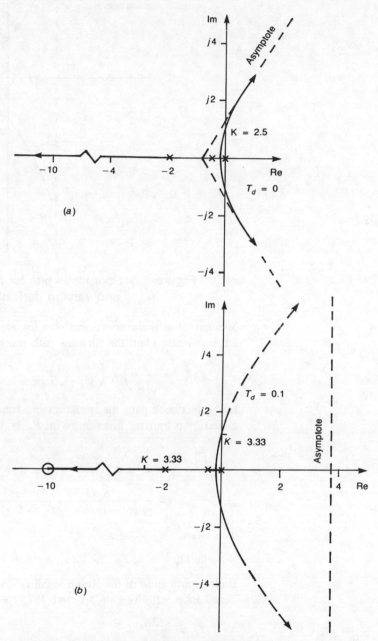

Figure 8.7 Effect of added derivative action on standard root locus plot

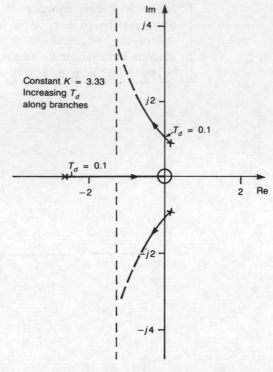

Constant K = 3.33
Increasing T_d
along branches

T_d = 0.1

T_d = 0.1

Figure 8.8 Root locus plot for fixed proportional gain
and varying derivative action

Solution For the proportional plus integral action controller in series with
the first order plant the forward path transfer function is

$$K_p \left(1 + \frac{1}{T_i s}\right) \quad \left(\frac{1}{1 + T_1 s}\right)$$

In the feedback path the measurement transfer function is $1/(1 + T_2 s)$. The
closed loop transfer function, with $K_p = 1$, $T_1 = 60$, $T_2 = 15$, is thus

$$\text{CLTF} = \frac{g(s)}{1 + g(s)h(s)}$$

$$= \frac{K_p(1 + T_i s)(1 + 15s)}{T_i s(1 + 15s)(1 + 60s) + K_p(1 + T_i s)}$$

with characteristic equation

$$900 T_i s^3 + 75 T_i s^2 + T_i(1 + K_p)s + K_p = 0$$

Direct application of the Routh stability criterion leads to the conditions for
closed loop stability as $K > 0$ and $75 T_i(1 + K_p) > 900 K_p$, i.e.

$$T_i > \frac{12 K_p}{1 + K_p}$$

Setting $T_i = 6$ s the root locus plot has an open loop zero at $-1/6$ and poles

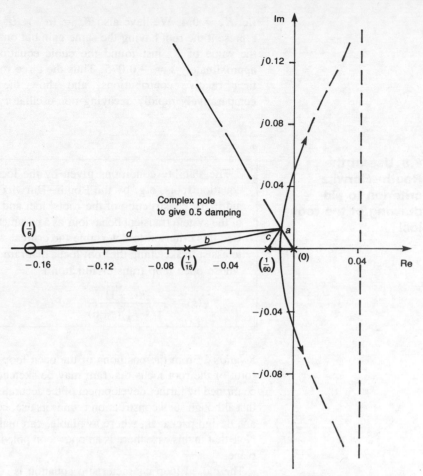

Complex pole
to give 0.5 damping

Figure 8.9 Root locus plot for system with integral
action

at 0, $-1/15$, $-1/60$ and critical gain $K_p = 1$. The root locus plot is shown
in Fig. 8.9 using the construction rules based on the angle condition.

For a damping factor of 0.5 in the quadratic factor in the closed loop poles,
the complex pair are given by where the line in the figure of slope $\phi = \cos^{-1} 0.5$ intersects the branch. The value of K may be determined by the
magnitude condition and used in turn to evaluate the third pole on the branch
between $s = -1/15$ and $s = -1/6$. From the figure there are closed loop
poles at approximately $-0.008 \pm j0.012$ and the gain to give these is given
by the value

$$\frac{K_p}{900 \times 6} = \frac{a \times b \times c}{d}$$

$$= \frac{0.014 \times 0.06 \times 0.014}{0.16} = 7.35 \times 10^{-5}$$

i.e. $K_p = 0.4$. We have also $K_p = (a' \times b' \times c')/d'$ where the a' etc. represent the root having the same gain but on the pole−zero branch. Using the value of K_p just found the cubic equation has a further real root at approximately $s = -0.075$. Thus the three roots have been found, but not their relative contributions, and show the oscillatory motion with a comparatively rapidly decaying non-oscillatory contribution also.

8.8 Use of the Routh−Hurwitz criterion to aid drawing of the root loci

> The stability conditions given by the location of the characteristic equation roots, e.g. by the Routh−Hurwitz criterion, may be used to aid in the construction of the roots' loci and hence in rapid assessment of the system transient behaviour as a parameter such as gain is changed. Use this criterion to determine the critical value of K for stability and to assist in sketching the root locus diagram for the closed loop system whose open loop transfer function is
>
> $$g(s) = \frac{k(s+1)}{s(s-1)(s+6)}$$

Solution From the positions of the open loop poles and zeros the probable form of the root locus diagram may be sketched. The shape of this may be confirmed by further development of the accurate plot. (Once again it is stressed that although such construction retains its use, computer root finding packages and design packages, where available, can make this much easier to solve.) Note that in this case there is an open loop pole in the right half of the complex plane.

The closed loop characteristic equation is

$$s(s-1)(s+6) + K(s+1) = 0$$

The Routh criterion gives the conditions, by the standard procedure used earlier, that $5(K-6) - K > 0$ and that $K > 0$. The first of these rearranges to $K > 7.5$, i.e. the critical value of gain is now a minimum value. From the root locus plot it is therefore expected that a branch (or branches) which is initially in the right half of the plane, with positive real part and therefore unstable pole, will move at increasing gain K into the left half plane. At the critical value of gain the characteristic equation must have a pair of conjugate imaginary roots lying on the imaginary axis. That is, $s = j\omega$ satisfies the equation

$$j\omega(j\omega - 1)(j\omega+6) + K(j\omega+1) = 0$$

i.e.

$$j\omega(-\omega^2 - 6 + K) - 5\omega^2 + K = 0$$

so that equating both real and imaginary parts of this equation to zero with $K = 7.5$ gives $\omega = 0$ and $\omega = \pm\sqrt{1.5}$, these being the crossing points on the imaginary axis. The subsequent root locus is as shown in Fig. 8.10.

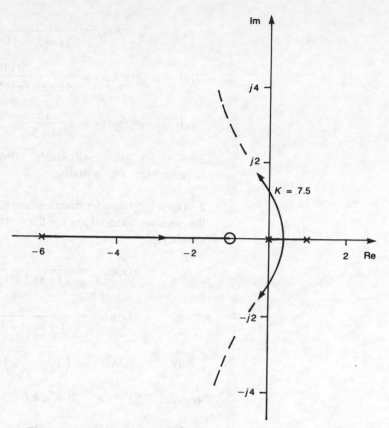

Figure 8.10 Root locus plot aided by Routh–Hurwitz results

Problems

Note. As with other graphical methods many of these problem solutions may be obtained or confirmed by use of computer packages or personal programs written to solve the closed loop equations.

1 The open loop transfer functions of systems with unity negative feedback are given below. Using the Routh–Hurwitz criterion, or by inspection, determine the closed loop stability/instability in each case.

(i) $\qquad g(s)h(s) = \dfrac{1}{1+0.5s}$

(ii) $\qquad g(s)h(s) = \dfrac{1}{s^2+3s+4}$

(iii) $\qquad g(s)h(s) = \dfrac{1}{(1+0.4s)(1+s+s^2)}$

(iv) $\quad g(s)h(s) = \dfrac{1}{(1-0.9s)(1+s+s^2)}$

(v) $\quad g(s)h(s) = \dfrac{20+0.3s}{(1+s)(1+0.5s)(1+3s)}$

(vi) $\quad g(s)h(s) = \dfrac{1}{s^2(1+3s)}$

Answer (i) Stable, (ii) stable, (iii) stable, (iv) unstable,
(v) unstable, (vi) unstable

2 Open loop transfer functions are given below. In each case what are the positive values (if any) of the variable parameter to give closed loop stability?

(i) $\quad g(s)h(s) = \dfrac{K}{(1+s)(1+0.5s)(1+3s)}$

(ii) $\quad g(s)h(s) = \dfrac{K}{(1+3s)(1+3s+5s)}$

(iii) $\quad g(s)h(s) = \left(1+\dfrac{1}{T_i s}\right)\dfrac{1}{(1+4s)(1+3s^2)}$

Answer (i) $K<14$, (ii) $K<4.6$, (iii) $T_i>6/7$

3 For each closed loop characteristic equation determine the range of positive K for closed loop stability:

(i) $\quad 2s^4 + s^3 + 2s^2 + s + K = 0$
(ii) $\quad s^4 + 3s^3 + s^2 + 2s + K = 0$

Answer (i) Unstable for all K, (ii) $K<2/9$

4 A closed loop system has the open loop transfer function

$$G(s) = \dfrac{K}{(1+0.3s)(1+0.7s)}$$

Draw the root locus plot. Describe qualitatively the closed loop behaviour as the gain K is increased and confirm the predictions by evaluating the closed loop response to a step input when $K = 0.1$ and $K = 1$.

Answer $0.0909(1+1.315e^{-3.037t}-2.315e^{-1.725t})$
$0.5[1-1.572e^{-2.381t}\ \sin(1.963t+0.6896)]$

5 Draw the closed loop root locus plot for the system with open loop transfer function

$$G(s) = \dfrac{K}{(1+s)(1+0.2s)(1+0.4s)}$$

What is the critical value of K for closed loop stability? At this value of K what is the frequency of oscillation of the closed loop system?

Answer $K = 12.6$, 4.472 rad s^{-1} ($\sqrt{20}$)

6 Sketch the root locus diagrams for the open loop transfer functions

(i) $\qquad G(s) = \dfrac{K(1+4s)}{(1+s)(1+0.5s)}$

(ii) $\qquad G(s) = \dfrac{K(1+0.25s)}{(1+s)(1+0.5s)}$

Describe the closed loop behaviour in each case as the gain K is increased progressively from a low to a high value.

Answer (i) Always overdamped. (ii) Overdamped for $K < 0.202$ and $K > 19.80$. Underdamped in middle range

7 Draw the closed loop root locus plot when the open loop transfer function is

$$G(s) = \frac{K}{(1+0.5s)(1+0.25s)^2}$$

What value of K will give closed loop roots with a damping coefficient of 0.7? At what value of K is closed loop instability reached?

Answer $K \simeq 0.5$, $K = 9$

8 A root locus plot has plant open loop poles at -1, -10, and -50. A 'phase advance compensator' is added to an original proportional controller, K. The plant and controller transfer functions are then respectively

$$g(s) = \frac{7500}{(s+1)(s+10)(s+50)} \quad \text{and} \quad k(s) = \frac{K(1+0.2s)}{1+0.025s}$$

(i) What is the critical value of K for closed loop stability for (a) the proportional controller and (b) the extended controller?

(ii) Show the effect on the closed loop root locus plot of the addition of the compensator and discuss it in terms of the transient response of the closed loop system.

(iii) Plot the Bode plot for each case using $K = 1$ and discuss these in relation to (i) and (ii).

Answer (i) 4.49, 5.90

9 Determine the range of positive gain K (if any) for closed loop stability when the open loop transfer functions are given by the following:

(i) $\qquad G(s) = \dfrac{K(1-s)}{s(1+s)}, \quad G(s) = \dfrac{K(s-1)}{s(1+s)}$

(ii) $$G(s) = \frac{K(1-0.4s)}{(1+0.5s)(1+s)}$$

(iii) $$G(s) = \frac{K(1+0.2s)}{(1-0.5s)(1+s)}$$

(iv) $$G(s) = \frac{K(1+0.4s)}{(1+0.5s)(s-1)}$$

Sketch also the corresponding Nyquist plots and show the root locus plots. What are the particular problems associated with right half plane poles and zeros? What is the limitation of the root locus diagram in dealing with these?

Answer   (i) $K<1$, unstable for all K,   (ii) $K<3.75$,
(iii) unstable for all K,   (iv) $K>1$

10   A closed loop negative feedback system has the open loop transfer function

$$G(s) = \frac{K}{(1+0.5s)(1+0.5s+0.25s^2)}$$

Draw the root locus diagram as K is varied. What is the critical value of K?

The addition of extra damping in this system has the effect of increasing the coefficient of s in the quadratic factor in the open loop transfer function $G(s)$ so that it becomes critically damped, i.e. has two equal first order factors. Show the new open loop poles on the diagram and sketch the resulting root locus, comparing the critical value of gain K and the transient response of this new closed loop system with that of the original as K is varied over its positive range.

Answer $K = 3$, $K = 8$

11   Draw the root locus plot for

$$G(s) = \frac{K}{(1+0.5s)(1+2s)(1+0.75s+0.25s^2)}$$

Show the effect on the plot of adding an open loop zero term $1+0.3s$ in $G(s)$. What is the critical value of K in each case for closed loop stability?

Answer $K = 5.0$, $K = 7.5$

9

Stability and the frequency domain

To extend the use of 'frequency response' plots beyond that of simply representing the response of a system to a particular input, one makes use of further criteria of stability and of system performance. A quantitative assessment of relative stability as well as absolute stability may be made, expressed in terms of 'phase margin' and 'gain margin'. Use is made of the polar plot and logarithmic plots introduced in Chapter 6. The comments already made about computer use very much apply here.

The Nyquist criterion

Once again this stability criterion is based upon locating the closed loop poles. By considering the mapping of a specific Nyquist path, Fig. 9.1, from the complex plane ($s = \sigma + j\omega$) into the complex plane of the open loop transfer function $G(s)H(s)$ the presence and number of unstable closed loop poles is determined.

The Nyquist stability criterion may be expressed in the following terms:

> For a stable closed loop system the Nyquist plot of $G(s)H(s)$ should encircle the $(-1, j0)$ point as many times N as there are poles of

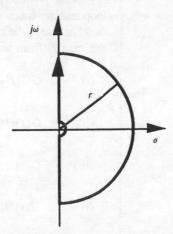

Figure 9.1 The Nyquist path in the complex plane

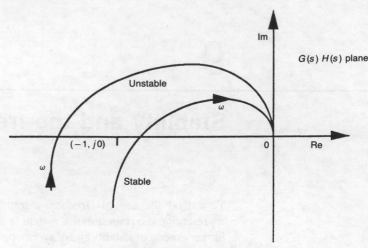

Figure 9.2 Basic Nyquist plots for closed loop stable and unstable systems

$G(s)H(s)$ in the right half of the s-plane. The encirclements (with the Nyquist path in the s-plane taken in a clockwise direction) will be in a counterclockwise direction. If the open loop function $G(s)H(s)$ is itself stable the critical point $(-1, j0)$ will not be enclosed.

This may be expressed in the following way. If there are P open loop poles in the right half plane, and hence encircled by the Nyquist path, then the number of closed loop poles in the right half plane will be Z where

$$Z = P + N$$

Thus if $P = 0$, open loop stable, then for closed loop stability $Z = 0$ and N should thus be zero also. If $P > 0$ then for closed loop stability $N = -P$. Stability and instability with the Nyquist plot are illustrated in Fig. 9.2.

9.1 Pattern of the Nyquist plots

Plot the Nyquist plots for the system shown in Fig. 9.3 with the open loop transfer functions $G(s)H(s)$ given below. Hence deduce the stability or otherwise of the closed loop system and the possible effects of changing the gain K from unity in each case.

(i) $\quad G(s)H(s) = \dfrac{K}{(0.5+s)(2+s)(1+s)} \qquad K = 1$

(ii) $\quad G(s)H(s) = \dfrac{K}{s(1+s)} \qquad K = 1$

(iii) $\quad G(s)H(s) = \dfrac{K}{s^2(1+s)} \qquad K = 1$

(iv) $\quad G(s)H(s) = \dfrac{K}{s(s^2+2s+2)} \qquad K = 1$

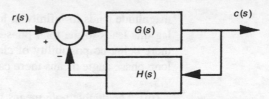

Figure 9.3 System for examples for analysis using the Nyquist criterion

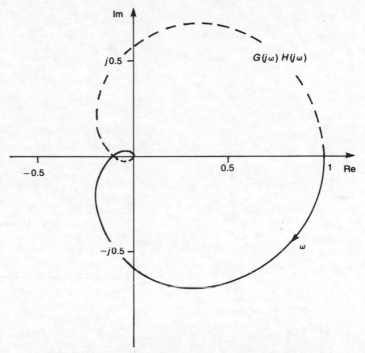

Figure 9.4 Nyquist plot for third order system

Solution For each of the open loop transfer functions the Nyquist plots are given. Only in the case where there is no pole at the origin, (i), is the full plot in the finite plane. However, where the low frequency plots move in from 'infinity' it is necessary to know the direction of the outer part of the plot in order to determine the enclosure or otherwise of the critical $(-1, j0)$ point. In each case the negative frequency range is a reflection in the real axis of the positive frequency section of the plot.

(i) This is a third order system with no zeros. The plot starts at $\omega = 0$ with unity magnitude and zero phase angle. As $\omega \to \infty$ the magnitude tends to zero and the phase angle to a total phase lag of 270°, Fig. 9.4. The magnitude at the phase lag of 180° is very small, there is no enclosure of the $(-1, j0)$ point and the system is closed loop stable. However, a large increase in the open loop gain above the given value of unity will eventually lead to instability, shown by enclosure of the $(-1, j0)$ point.

(ii) The pole at the origin gives a 90° phase lag at all frequencies and the magnitude tends to infinity at low frequency. Combined with the first order lag this leads to the total phase angle lying between $-90°$ and $-180°$ and there being no possibility of closed loop instability as the range of the open loop phase angle means there can be no enclosure of the $(-1, j0)$ point, Fig. 9.5.

(iii) The double pole means that there is a minimum of $180°$ of phase lag.

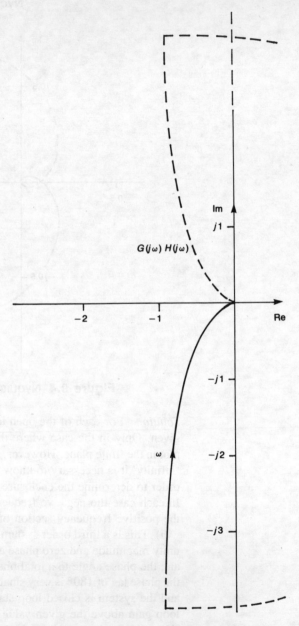

Figure 9.5 Nyquist plot for second order system with pure integrator

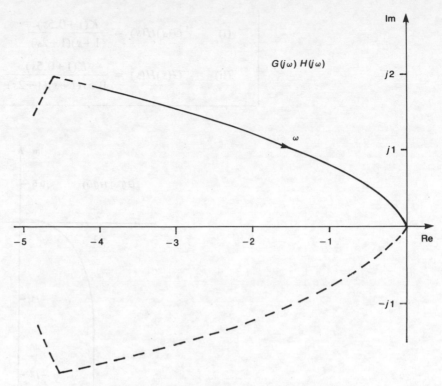

Figure 9.6 Nyquist plot with double integrator

Thus the positive frequency plot always lies above the real axis. The principle of conformal mapping means that the small indentation about the origin in the Nyquist path maps into a part of the plot which comes up from below the real axis as shown, Fig. 9.6. Thus the full Nyquist plot, the full mapping of the Nyquist path, gives two encirclements in the clockwise sense of the $(-1, j0)$ point and the closed loop system in unstable for all values of gain greater or less than unity.

(iv) This system is again third order, having an open loop phase lag ranging from $90°$ to $270°$, Fig. 9.7. With the gain $K = 1$ the closed loop system is stable but increasing the gain will lead to encirclement of the critical point and an unstable closed loop. Lower values of K keep the closed loop system stable.

9.2 Systems with zeros

Similarly to the above question, plot the full Nyquist plots for the systems with the open loop transfer functions $G(s)H(s)$ given below. These functions are of the form produced by introducing derivative and integral action control respectively to a second order system. Again deduce the stability or otherwise of the closed loop system with $K = 1$ and the possible effects of changing this gain.

$$\text{(i)} \qquad G(s)H(s) = \frac{K(1+0.5s)}{(1+s)(1+2s)}$$

$$\text{(ii)} \qquad G(s)H(s) = \frac{K(1+0.5s)}{0.5s(1+s)(1+2s)}$$

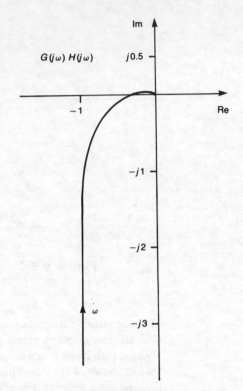

Figure 9.7 Nyquist plot for third order system with pure integrator

Solution   (i) The addition of derivative action adds phase advance to the system. The net phase lag is thus reduced and the phase lag at high frequencies is 90° less than it would be without it. In this case the positive frequency section comes into the origin tangentially to the negative imaginary axis, i.e. at −90°, Fig. 9.8. The closed loop system is stable for all K.

(ii) Integral action adds both a pole at the origin and a zero at $-1/T_i$. The low frequency phase angle is thus affected by the continuous addition of the additional 90° lag from the pole while at the higher frequency the net phase contribution is zero from the pole−zero combination. The phase lag exceeds 180° during the plot and high gain values for K will lead to closed loop instability, Fig. 9.9.

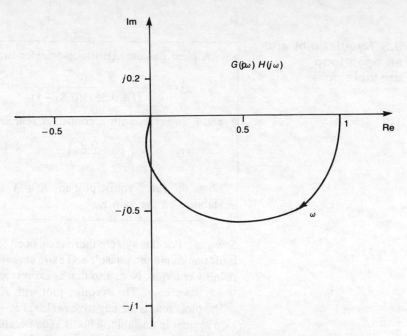

Figure 9.8 Nyquist plot for system with derivative action

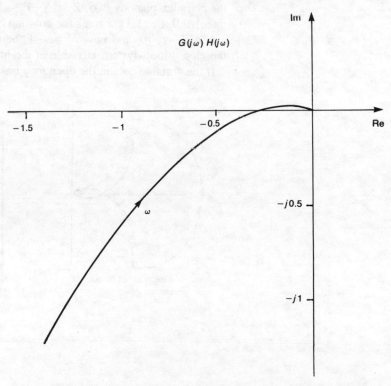

Figure 9.9 Nyquist plot for system with integral action

**9.3 Nyquist plot and
an open loop
unstable pole**

A plant has the open loop transfer function

$$g(s) = \frac{1}{(1+0.25s)(0.5s-1)}$$

and is in series with a controller represented by

$$k(s) = K\left(1 + \frac{0.2}{s}\right)$$

Show the full Nyquist plot for $K = 1$ and explain the dependence of stability on the gain K.

Solution For this system there is an open loop unstable pole, i.e. the system is of 'non-minimum phase', and extra care is required in the use of the Nyquist stability criterion. Note also that as expressed the transfer function has negative steady state gain. The Nyquist plot with $K = 1$ is as shown in Fig. 9.10.

The plot crosses the negative real axis at -0.9. Thus with $K = 1$ the closed loop system is unstable. This is seen because there is an unstable open loop pole, $P = 1$, and there is one clockwise encirclement of the critical point so that $N = 1$ also. Thus the number of closed loop poles in the right half of the complex planc is two, $Z = N+P = 2$. Increasing the gain K to greater than $1/0.9$, i.e. 1.11, brings the crossing of the real axis to the more negative side of $(-1, j0)$ and now $N = -1$, being anticlockwise. Now $Z = 0$ and the closed loop system is stable at the higher values of gain.

If the unstable pole in the open loop transfer function had been $1/(1-0.5s)$

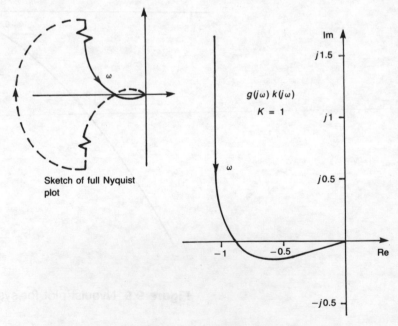

Sketch of full Nyquist plot

Figure 9.10 Effect of unstable open loop pole

then the total phase angle would be advanced by 180°, the full plot would rotate to fall fully in the right half plane and the encirclements, whatever the value of gain K, would be zero. Thus $Z = 1$ and the closed loop system is always unstable. (These results may be confirmed using the Routh–Hurwitz criterion.)

Relative stability

While the Nyquist stability criterion gives a ruling on a system's absolute closed loop stability, the 'closeness' of a system to instability may be expressed in terms of the stability margins, i.e. the phase margin and the gain margin. These may be shown graphically on the Nyquist plot or evaluated numerically from the defining equations of the system behaviour. The phase margin is the amount by which the open loop phase lag may increase at the gain crossover frequency (that frequency at which the open loop gain is unity) before instability is reached. The gain margin is the factor by which the open loop gain may be increased before instability is reached. It is the reciprocal of the open loop gain, $|G(s)H(s)|$, at the phase crossover frequency (that frequency at which the open loop phase lag reaches 180°). This gain margin is normally expressed in decibels. Figure 9.11 illustrates the derivation of the stability margins from the Nyquist plot.

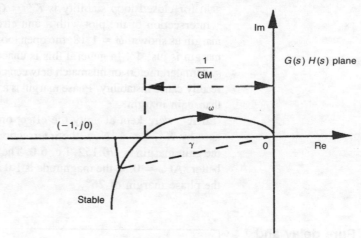

Figure 9.11 Gain and phase margins (GM and γ)

9.4 Gain and phase margins

Taking the open loop system

$$G(s)H(s) = \frac{K'(1+0.5s)}{s(1+s)(1+2s)}$$

determine the closed loop gain and phase margins when $K' = 4$ and $K' = 1$.

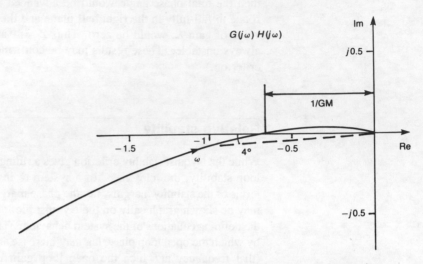

$G(j\omega)\, H(j\omega)$

Figure 9.12 Determination of phase and gain margins

Solution This is the same dynamic system as used in Example 9.2(ii) above. Repeating that solution but with $K' = 4$ it is seen that the plot, Fig. 9.12, cuts the real axis at -0.66 when the open loop phase is $-180°$ at a frequency of $\omega = 1.45$ (rad s^{-1}). The gain margin is $1/0.66$, i.e. **1.52**. The limiting gain for closed loop stability is $K' = 6.0$. [$K' = 2K$ for K in 9.2(ii)].

Intersection of the plot with a unit circle about the origin gives the phase margin as shown, $\omega = 1.18$; the open loop phase angle is $-176°$ so the phase margin is just **4°**. In general this is unacceptably low as a small deviation in system parameters or mismatch between control design values and reality could readily cause instability. Phase margin is a better indicator of the system stability than gain margin.

If K' were kept at unity the effect on the plot is the same as scaling the axes by K', i.e. 4. Now the intersection on the real axis is at -0.152 and the gain margin is $1/0.152$, i.e. **6.0**. The phase margin becomes significantly better. At $\omega = 0.58$ the magnitude is 1.0 and the phase angle is $-154°$ giving the phase margin of **26°**.

9.5 Pure delay and the Nyquist plot

If the system described by the transfer function in Example 9.4 is modified such that a pure delay of 1 s is incurred in the feedback loop, what effect does this have on the closed loop stability?

Solution The effect of the pure delay is to add the term e^{-sT} to the open loop transfer function and a phase lag of ωT where T is the pure delay and ω the particular frequency. It reduces therefore the relative stability of the system and can rapidly cause instability at relatively low gain. In this particular case the system is closed loop unstable at both $K' = 4$ and $K' = 1$, the plot, Fig. 9.13, cutting the negative real axis beyond the $(-1, j0)$ critical point. The new value of gain for limiting closed loop stability is $K' = 4/4.7 = 0.85$.

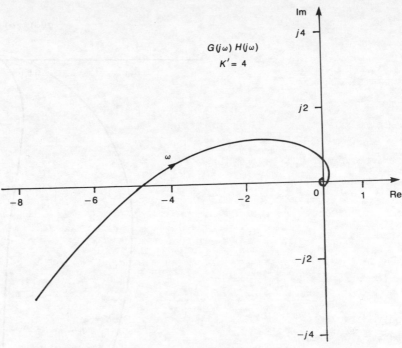

$$G(j\omega)\,H(j\omega)$$
$$K' = 4$$

Figure 9.13 Effect of pure delay on stability as shown by Nyquist plots

9.6 Effect of added derivative action

Additional controller terms have been seen in earlier examples to affect system dynamics and stability. A plant and controller with unity feedback have the combined open loop transfer function

$$G(s) = g(s)k(s) = \frac{9}{s(s^2+3s+9)} \cdot K(1+T_d s)$$

By use of the Nyquist plot determine the limiting value of K when T_d is set to zero and the gain and phase margins at this value of K when T_d is 0.2.

Solution The system is third order with a pure integrator and in the absence of the derivative control term has a phase angle of $-90°$ at $\omega \to 0$ and tends to $-270°$ as frequency increases, Fig. 9.14. Using $K = 1$ the phase angle of $-180°$ gives the intersect on the real axis at -0.333, a gain margin of 3.0 and hence limiting gain of **3.0**. With $K = 1.0$ the phase margin is $68°$.

The effect of adding the derivative action is to reduce the phase lag and improve the stability characteristics. In particular it is observed that at high frequency the net open loop phase angle now tends to $-180°$. However, the plot shows that instability is still possible at high gain, the gain margin being **2.5** (with the gain $K = 3$). The phase margin even at the higher gain of $K = 3$ is still now about $24°$.

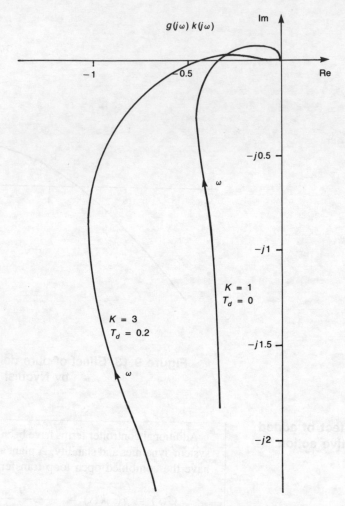

$g(j\omega)\,k(j\omega)$

Figure 9.14 Limiting gain and stability margin

Stability and the inverse Nyquist plot

The inverse polar plot, Fig. 9.15, may be more readily used in some cases as described earlier and it also has a wider use in multi-variable systems analysis. The stability criterion may be applied using this method where the mapping is now into the $[G(s)H(s)]^{-1}$ plane.

From the characteristic equation for the closed loop system, i.e.

$$F(s) = 1 + G(s)H(s) = 0$$

the alternative form of the relationship may be obtained such that

$$F'(s) = \frac{1}{G(s)H(s)} + 1 = 0$$

The zeros of this expression are still the closed loop poles. The poles of F'

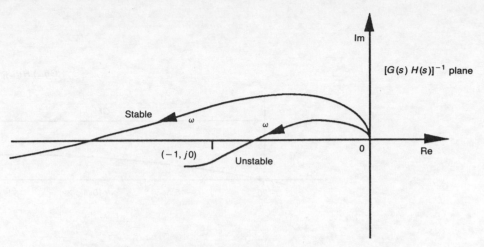

Figure label inside plot: Stable, ω, ω, (−1, j0), Unstable, 0, Re, Im, $[G(s)\,H(s)]^{-1}$ plane

Figure 9.15 Inverse Nyquist plots for simple closed loop
stable and unstable systems

are those of $[G(s)H(s)]^{-1}$, the zeros of $[G(s)H(s)]$. In terms of the inverse
plot the same stability criterion means that 'for a stable closed loop system
the net number of rotations of the mapping $F'(s)$ about the origin must be
counterclockwise and equal to the number of poles of $[G(s)H(s)]^{-1}$, i.e.
zeros of $[G(s)H(s)]$, that lie in the right half plane'.

9.7 Determination of closed loop gain and phase margins

Use the inverse Nyquist plot to determine the gain and phase margin
of the system with open loop transfer function

$$G(s)H(s) = \frac{0.3(1+0.5s)}{s(1+s)(1+2s)}$$

Solution This system, Fig. 9.16, is the same as an earlier Nyquist plot
(Example 9.2), but with $K = 0.15$, and gives the opportunity to compare both
the general shape and the use of the straightforward stability criterion

$$[G(j\omega)H(j\omega)]^{-1} = \frac{j\omega(1+j\omega)(1+2j\omega)}{0.3(1+0.5j\omega)}$$

so that the magnitude and argument are respectively

$$\frac{\omega\sqrt{(1+\omega^2)}\sqrt{(1+4\omega^2)}}{0.3\sqrt{(1+0.25\omega^2)}}$$

and

$$90° + \tan^{-1}\omega + \tan^{-1}2\omega - \tan^{-1}0.5\omega$$

Now at low frequency the plot has small magnitude and a phase angle of 90°,
reaching 180° at the high frequency values and exceeding this value so that
the real axis is cut by the plot. The gain margin is shown, having a value of

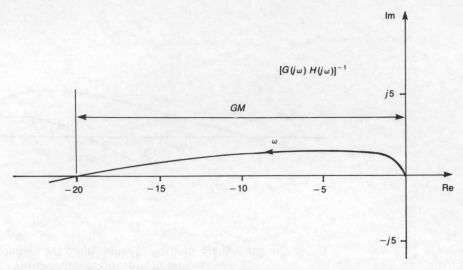

$[G(j\omega)\,H(j\omega)]^{-1}$

$j5$

GM

ω

-20 -15 -10 -5 Re

$-j5$

Figure 9.16 Stability margins from the inverse Nyquist
plot

20. The critical gain value is thus 0.15×20, i.e. 3.0. This confirms the previous finding using the Nyquist plot. The phase angle at which the plot cuts the unit radius, i.e. at which there is unit magnitude, is 125°. The phase margin is **55°** $(180° - 125°)$.

9.8 Open loop unstable system

Derive, using the inverse Nyquist plot, the limiting range of K for stability of the closed loop when the open loop transfer function is

$$G(s)H(s) = \frac{1}{(1+0.25s)(0.5s-1)} \cdot K\left(1+\frac{0.2}{s}\right)$$

Solution For the more contrived transfer function shown with an open loop unstable pole

$$[G(j\omega)H(j\omega)]^{-1} = \frac{j\omega(1+0.25j\omega)(0.5j\omega-1)}{K(j\omega+0.2)}$$

The inverse Nyquist plot, with K equal to unity, for this system is given in Fig. 9.17. Application of the stability criterion to this plot, confirmed by a standard plot, Fig. 9.10, or the Routh–Hurwitz method, shows the closed loop system to be stable for gains greater than $K = 1.1$. The plot in Fig. 9.17 as drawn for $K = 1$ shows an unstable closed loop system.

9.9 Direct comparison of direct and inverse Nyquist plots

Use the direct Nyquist and inverse Nyquist plots for the system with open loop transfer function

$$G(s) = \frac{K(s-1)}{(s+2)(s+3)}$$

to show the limitations arising from systems having open loop zeros in the 'right half plane'.

Solution The direct and inverse Nyquist plots are shown in Figs 9.18 and 9.19 for $K = 1$. Applying the criterion to each of these gives a limiting value on the gain for stability of $K < 6$. (The Routh criterion and root locus methods when applied also indicate this result.)

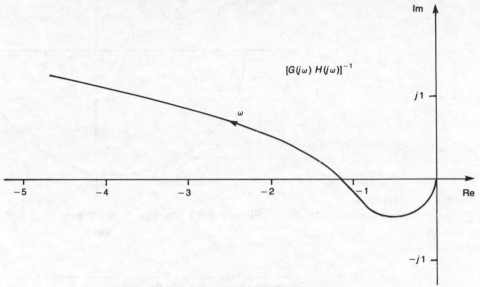

Figure 9.17 Determination of critical gain for system with unstable open loop pole

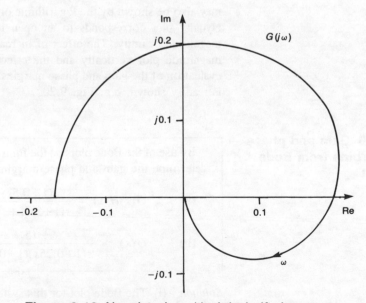

Figure 9.18 Nyquist plot with right half plane zero

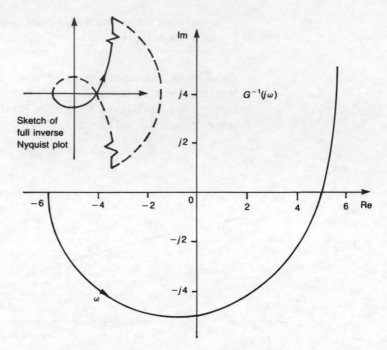

Figure 9.19 Inverse Nyquist plot with right half plane zero

Stability and the Bode plot

As was seen earlier the information which is conveyed by the Nyquist plot may also be shown by the logarithmic or Bode plot. The critical point in the Nyquist plot corresponds to an open loop phase angle of $-180°$ and a magnitude of unity. The effect of increasing open loop gain is to move the magnitude plot vertically and the effect of changing gain, and hence the evaluation of the gain and phase margins and of the critical gain for stability, is readily shown, e.g. Fig. 9.20.

9.10 Gain and phase margins from Bode plot

By use of the Bode plots on the following open loop transfer functions determine the gain and phase margins:

(i) $$G(s)H(s) = \frac{4(1+0.5s)}{s(1+s)(1+2s)}$$

(ii) $$G(s) = \frac{15}{s(1+0.25s)(1+0.05s)}$$

Solution   (i) The Bode plot for this system, Fig. 9.21, is rapidly built up by use of the magnitude corner frequency, or asymptotic, plots for each system

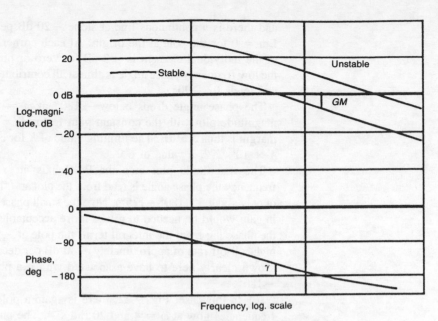

Figure 9.20 Stability on the Bode plot

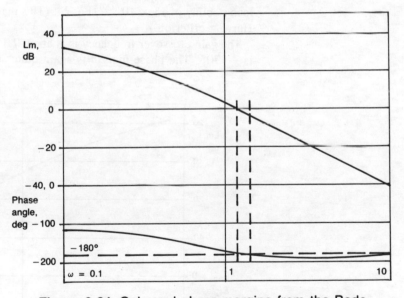

Figure 9.21 Gain and phase margins from the Bode plot

element. The total phase is the sum of that of the individual elements (although less accurate values may be obtained by other straight line segments).

The corner frequencies are

for the zero, $\omega = 2(\text{rad s}^{-1})$
for the poles, $\omega = 0.5, 0.1$

and there is a continuous line at slope -20 dB per decade through $\omega = 1$, Lm $= 0$ for the pole at the origin. At each corner frequency the magnitude of the individual element is ± 3 dB for zero or pole respectively. Note that the low frequency phase angle is almost all contributed by the pole at the origin and tends to $-90°$.

The phase angle drops below $-180°$ at $\omega = 1.4$ rad s^{-1} at which the magnitude plot with the constant gain factor $K = 4$ is -3.5 dB. The gain margin is thus **3.5 dB**. The limiting value of K for closed loop stability is thus $4 \times 10^{(3.5/20)}$, a value of 6.0.

The magnitude plot cuts the 0 dB axis at $\omega = 1.2$ rad s^{-1}. At this frequency the phase angle is read from the plot at $-175°$ giving a phase margin of only about **5°** ($180° - 175°$). Note the small phase margin. A large reduction in gain would be needed to give a more acceptable phase margin because of the phase lag from the integral term, the pole at -0.1 s^{-1}, and the resulting high roll-off rate of approximately -40 dB per decade on the magnitude plot, shown clearly here to have associated with it a phase angle of the order of $-180°$.

(ii) In this case, Fig. 9.22, there is again a pole at the origin and corner frequencies now at $\omega = 4$ and 20 rad s^{-1}. The phase angle reaches $-180°$ at $\omega = 9$ rad s^{-1} at which the magnitude value is -4 dB. The gain margin is thus **4 dB**. This gives a limiting value of the constant gain factor for closed loop stability of $15 \times 10^{(4/20)}$, i.e. 24. (This may be confirmed by the Routh–Hurwitz criterion.)

The gain crossover frequency is at about 7.1 rad s^{-1} when the phase angle is $-170°$. The phase margin is again quite small at **10°**.

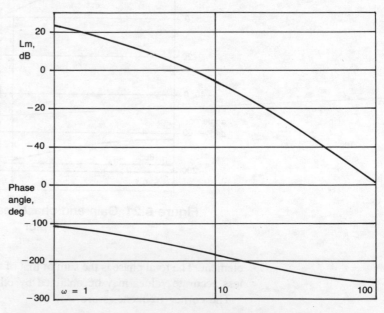

Figure 9.22 Bode plot requiring higher frequency range

9.11 Second order factors and pure delay

Show the effect on the Bode plot, closed loop stability, and critical gain value K_{crit} of the addition of derivative action and of a pure delay by plotting the Bode plots for

(i) $\qquad G(s) = \dfrac{9}{s(s^2+3s+9)} \cdot K(1+T_ds)$

(ii) $\qquad G(s) = \dfrac{9e^{-s}}{s(s^2+3s+9)} \cdot K(1+T_ds)$

each for the two cases with $K = 2$ but with $T_d = 0$ and $T_d = 0.2$ respectively.

Solution The effect of the pure delay, which adds a phase lag of ω but no additional gain (it has a magnitude factor of unity for all frequencies), is shown in the following plots. All subsequent numerical values are evaluated from these plots.

(i) *Without the additional phase lag* the critical gain for closed loop stability is significantly affected by the use of derivative action:

(a) Without derivative action the gain crossover frequency is 2.35 rad s^{-1} with corresponding phase angle of $-153°$, a phase margin of **27°**, Fig. 9.23. When the phase angle reaches $-180°$ at $\omega = 3.0$ rad s^{-1} the gain margin is **3.6 dB**. The critical value of K is $K_{crit} = 2 \times 10^{(3.6/20)} = 3.0$.

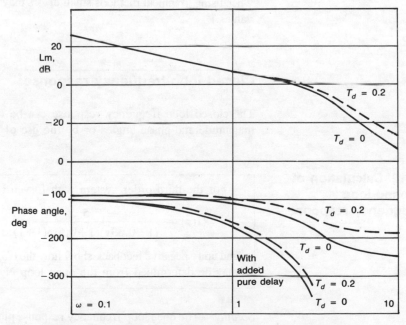

Figure 9.23 Effect on Bode plot of added derivative action and pure delay

(b) On adding the derivative action the gain crossover changes slightly to 2.6 rad s^{-1} and the phase margin improves to **45°** $(180° - 135°)$. The gain margin (at $\omega = 4.9$ rad s^{-1}) has increased to about **12 dB**. The critical gain value will be $K_{crit} = 2 \times 10^{(12/20)} = \mathbf{8.0}$. Both in terms of phase margin and allowable gain the closed loop system has been significantly improved.

(ii) With the delay added, e.g. as in some measurements, the magnitude plot is unaffected as the delay introduces no additional gain factor other than unity. However, there is a major change in each of the phase angle plots as the frequency increases.

(a) Without the proportional controller, $T_d = 0$, the phase reaches $-180°$ at $\omega = 1.13$ rad s^{-1} at which the magnitude plot has the value of about 5.7 dB. This is a 'negative phase margin', i.e. the system is unstable at gain $K = 2$. To reach the marginally stable condition K must be *reduced* by 5.7 dB. This amount of -5.7 dB is a factor of $10^{(-5.7/20)}$ which is 0.5. Thus the gain must be reduced to the new critical value $K_{crit} = \mathbf{1}$.

(b) With the addition of the derivative control term the phase crossover at $-180°$ is at $\omega = 1.33$ rad s^{-1}. The magnitude is 4.7 dB and again the closed loop system is unstable. The gain K must be reduced by the factor $10^{(-4.7/20)}$ to give a critical value of $K_{crit} = \mathbf{1.2}$.

Although the derivative action is beneficial in both cases in reducing the phase lag the effect of the pure delay is overriding, producing an unstable system with both controllers unless there is a major reduction in gain K. (Note that when using graphical methods small errors may arise when reading off specific values.)

Closed loop frequency response

The closed loop frequency response can be calculated from its open loop magnitude and phase angles or by the use of further graphical extensions.

9.12 Calculation of closed loop frequency response

> For the third order system with forward path transfer function
>
> $$G(s) = \frac{5}{(1+0.33s)(1+0.5s)(1+s)}$$
>
> and unity negative feedback show how the *closed loop* frequency response may be determined from the open loop Nyquist plot.

Solution The open loop frequency response plot, the Nyquist plot for positive frequency, can be used as a basis for measurements from which to construct the *closed loop* frequency response. Note that the term 'Nyquist' is reserved for the open loop plot. If the open loop frequency transfer function is $G(j\omega)$

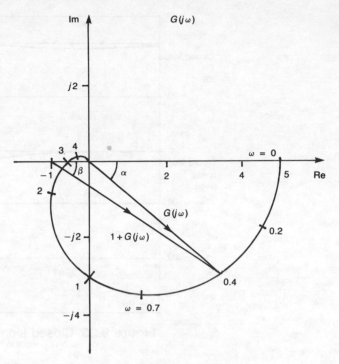

Figure 9.24 System Nyquist plot (open loop frequency response)

then, with unity feedback, the closed loop transfer function is $G(j\omega)/[1+G(j\omega)]$ with magnitude term $|G(j\omega)|/|1+G(j\omega)|$. The phase angle is $\arg[G(j\omega)] - \arg[1+G(j\omega)]$, i.e. $\alpha - \beta$ for the selected point on the open loop plot, Fig. 9.24.

The closed loop response calculated using values taken directly from this plot is shown in Fig. 9.25 as the separate magnitude and phase plots (of similar form to, but *not*, a Bode plot). In order to do this the frequency must appear as a parameter along the open loop polar plot.

Alternatively the closed loop response may be calculated directly using the above relationships for gain and phase. From Fig. 9.24 the closed loop magnitude at a frequency ω is

$$\frac{|G(j\omega)|}{|1+G(j\omega)|} = \frac{|G(j\omega)|}{\sqrt{[(1+|G(j\omega)|\cos\alpha)^2 + (|G(j\omega)|\sin\alpha)^2]}}$$

and the phase angle can be seen from Fig. 9.24 to be

$$\beta = \tan^{-1}\frac{|G(j\omega)|\sin\alpha}{1+|G(j\omega)|\cos\alpha}$$

where $\alpha = \arg(G(j\omega))$.

Note the low frequency gain tending to 5/6 (-1.6dB) and the high frequency gain tending to zero. Low frequency phase lag is zero also and tends to a high frequency value of 270°.

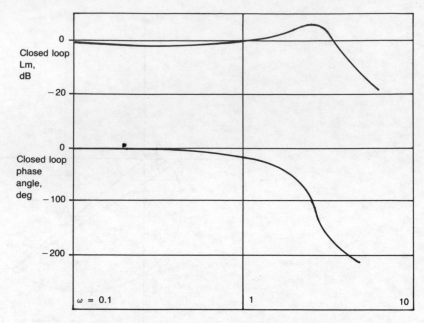

Figure 9.25 Closed loop frequency response

M and *N* circles and the Nichols chart

The closed loop frequency response of linear systems may also be derived in other ways directly from open loop graphical methods. This is achieved by superimposing in the complex plane lines of constant closed loop magnitude (*M* loci) and constant closed loop phase angles (α or *N* loci). Each of these form a family of circles and the intersections of the open loop plot (the Nyquist plot) with these give the corresponding closed loop frequency response.

If the open loop frequency function $G(j\omega)$ is expressed as

$$G(j\omega) = x + jy$$

then the closed loop magnitude *M* is given by the relationship

$$\left(x - \frac{M^2}{1-M^2}\right)^2 + y^2 = \left(\frac{M}{1-M^2}\right)^2$$

This family of circles is shown in Fig. 9.26.

The closed loop phase angle α is given by

$$\left(x + \frac{1}{2}\right)^2 + \left(y - \frac{1}{2N}\right)^2 = \frac{N^2+1}{4N^2}$$

where $N = \tan\alpha$. This corresponding set of circles is given in Fig. 9.27. [These circles are multi-values so that $N = \tan(\alpha + m180°)$.]

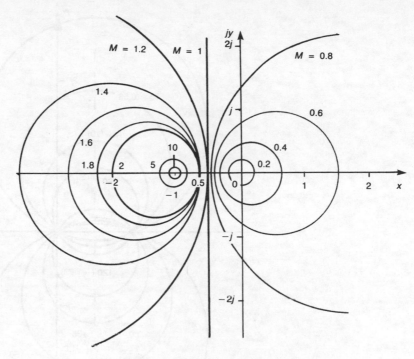

Figure 9.26 Constant *M* circles

The constant *M* and *N* circles are given in Figs 9.26 and 9.27 respectively. Show the superposition of the Nyquist plot for the open loop function $G(s)$ and sketch the closed loop frequency response if

$$G(s) = \frac{15}{s(1+0.25s)(1+0.05s)}$$

Solution Superimposing the Nyquist plot for positive frequencies on the *M* and *N* circles leads directly at the points of intersection to the closed loop response in terms of magnitude and phase angle. These values are prone to graphical errors from small plots but indicate the principal features of the closed loop frequency response and give ready comparison with open loop behaviour. The magnitude (*M*) and phase (*N*) are shown here plotted together on the same axes, Fig. 9.28, thus allowing closed loop magnitude and phase to be obtained if required from the single open loop plot. The open loop Nyquist plot has frequency as a parameter along its length and this is required to plot Fig. 9.29.

Note that the plot is essentially in cartesian, as distinct from polar, coordinates and is most easily plotted that way, the magnitude and phase pair of the frequency response being converted to real and imaginary parts. From the intersections of the *M* circles and the Nyquist plot the following closed loop frequency response curve is obtained.

With more extensive calculation, e.g. by the availability of a computer system

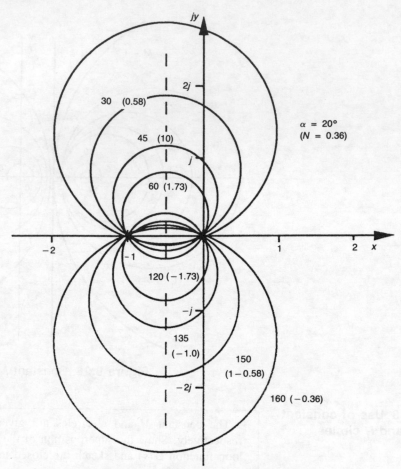

Figure 9.27 Constant *N* circles

program, a more accurate set of results is readily obtained and confirmation of the values is obtained also from the trigonometry of the relationship between $G(j\omega)$ and $1 + G(j\omega)$ in the complex plane as shown in Example 9.12. In the absence of the *M* and *N* plots or a Nichols chart this is an alternative way to proceed. For frequency points which may be taken from the diagram the corresponding magnitude values are as follows:

ω rad s^{-1}	(2)	4	7	9	10	12	15	20
M	(1.08)	1.40	5.47	1.61	0.99	0.50	0.25	0.11

Nichols chart

The Nichols chart is a compact equivalent of the constant *M* and *N* circles. The circle information is redrawn and superimposed on a log-magnitude *versus*

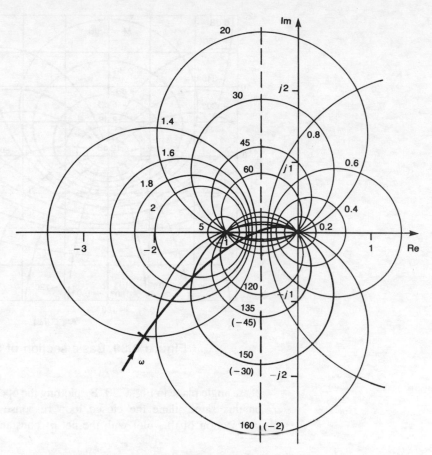

Figure 9.28 'Nyquist plot' superimposed on the *M* and *N* circles

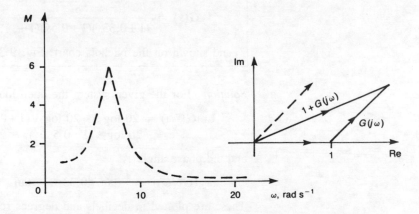

Figure 9.29 Derived closed loop frequency response

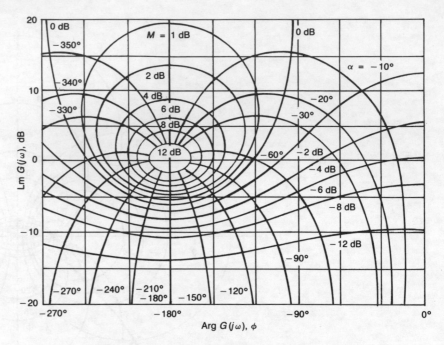

Figure 9.30 Basic section of Nichols chart

phase angle plane in Fig. 9.30. By plotting the open loop gain/phase relationship on this same plane the closed loop response may be determined by the intersection of this plot with the net of constant M and N loci.

9.14 Use of Nichols chart

> Show, by selecting discrete values of frequency, how the closed loop frequency response of the system defined by the open loop transfer function
>
> $$G(s) = \frac{5}{(1+0.33s)(1+0.5s)(1+s)}$$
>
> and shown on the Nichols chart, Fig. 9.31, can be predicted.

Solution For the given system the open loop log-magnitude is

$$\begin{aligned} \mathrm{Lm}G(j\omega) = {} & 20 \log 5 - 20 \log \sqrt{(1+0.33^2\omega^2)} \\ & - 20 \log \sqrt{(1+0.5^2\omega^2)} - 20 \log \sqrt{(1+\omega^2)} \end{aligned}$$

and the phase angle is

$$\arg(G(j\omega)) = -\tan^{-1}0.33\omega - \tan^{-1}0.5\omega - \tan^{-1}\omega$$

These are plotted in decibels and degrees respectively in Fig. 9.31.

Note that closed loop gain and phase margins may be determined from this form of open loop plot. It can be seen that when the open loop phase angle

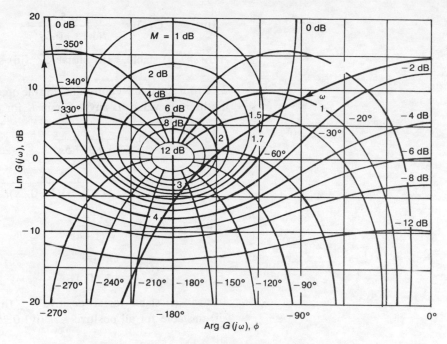

Figure 9.31 Closed loop gain values derived from the Nichols chart

is −180° the open loop log-magnitude is approximately −6 dB, giving a gain margin of 6 dB. Similarly the open loop phase angle at unity gain is approximately −155° giving a phase margin of 25°.

The closed loop magnitude displays a resonance peak before falling away quickly at the higher frequencies. At low frequencies the closed loop gain tends to 5/6, i.e. −1.6 dB and there is no phase lag between input and output. As the system used in this question is the same as that in Example 9.12 the derived closed loop characteristic is as in Fig. 9.25.

Problems

1 Draw the Nyquist plot for the following open loop transfer functions $G(s)$ using the positive frequency range. Indicate the full Nyquist plot and deduce whether or not the negative feedback closed loop system is stable.

(i) $G(s) = \dfrac{10}{1+0.5s}$

(ii) $G(s) = \dfrac{5}{s(1+0.5s)(1+0.8s)}$

(iii) $G(s) = \dfrac{0.5}{s(1+s+0.5s^2)}$

Answer   (i) stable,   (ii) unstable,   (iii) stable

2   From suitable Nyquist plots for the open loop systems below deduce the limiting value (if any) of gain K for closed loop stability:

(i) $G(s) = \dfrac{K}{1+3s+5s^2}$

(ii) $G(s) = \dfrac{K}{(1+0.5s)(1+0.8s)(1+s)}$

(iii) $G(s) = \dfrac{K(1+s)}{s^2(1+4s)}$

(iv) $G(s) = \dfrac{K\,e^{-0.2s}}{1+0.4s}$

Answer   (i) stable for all positive K,   (ii) 8.775,
(iii) unstable for all positive K,   (iv) 0.26

3   A dynamic system is open loop unstable with transfer function

$$g(s) = \dfrac{1}{s(1+0.3s)(0.5s-1)}$$

It is intended to control this with a simple proportional controller. Draw the Nyquist plot and assess the stability of the closed loop if only proportional action is used. What range, if any, of proportional gain K would be suitable?

Answer   Unstable for all K

4   The open loop description of a servo system is given by the transfer function

$$G(s) = \dfrac{1}{s(1+1.2s+s^2)}$$

(i) Use the Nyquist plot to determine the gain and phase margins of the closed loop system.
(ii) Check these stability margins by direct calculation of the gain and phase angles at the crossover frequencies.
(iii) Confirm these findings by use of a Bode plot.

Answer   1.2 (1.6 dB), 9.3°

5   To increase the phase margin of problem **4** a proportional plus derivative action controller is proposed:

$$k(s) = K(1+T_d s)$$

If $T_d = 2$ s what is the new phase margin if $K = 1$? If K is increased to the critical value indicated by **4** (i) above what now is the phase margin and new gain margin?

Show the effect of these changes on Nyquist plot sketches.

Answer 37.9°, 33.3°, ∞ (stable for all K)

6 Plot the inverse polar plot (inverse Nyquist) for the open loop transfer functions

(i) $G(s) = \dfrac{1+0.3s}{1+0.7s}$

(ii) $G(s) = \dfrac{K}{s(1+3s)(1+0.8s)}$ with $K = 1$

For system (ii) hence determine the limiting value of K for stability.

Answer $K = 1.58$

7 Plot the inverse polar plot for

$$g(s) = \frac{1}{(1+3s)(1+2s)}$$

Velocity feedback is used with feedback gain k_v. If $k_v = 2$ use the combination of inverse plots to give

$$\left[\frac{g(j\omega)}{1+g(j\omega)k_v j\omega} \right]^{-1}$$

8 Use the Nyquist and the inverse Nyquist plots to determine the range of gain K for the open loop system

$$G(s) = \frac{K(1.5s-1)}{(1+s)(1+0.5s)}$$

to be closed loop stable.

Answer $K < 1$

9 Use the Bode plots to determine the closed loop stability and, where appropriate, the gain and phase margins, for the following open loop systems:

(i) $G(s) = \dfrac{2}{1+3s+5s^2}$

(ii) $G(s) = \dfrac{6}{(1+0.5s)(1+0.8s)(1+s)}$

(iii) $G(s) = \dfrac{1(1+s)}{s^2(1+4s)}$

(iv) $G(s) = \dfrac{1.5\, e^{-0.2s}}{1+0.4s}$

Answer (i) 65.7°, ∞; (ii) 13.6°, 1.46 (3.3 dB);
(iii) unstable for all gain values; (iv) 99.8°, 2.54 (8.1 dB)

10 The open loop system

$$G(s) = \dfrac{K}{(1+0.7s)(1+0.8s+0.2s^2)}$$

is to be closed loop stable with a phase margin of 50°.
   (i) Use the Bode plot to determine the required value of K.
   (ii) Using the gain crossover frequency at this value of K confirm the phase margin by calculation.

Answer (i) 2.46

11 A third order system has the open loop transfer function

$$G(s) = \dfrac{K(s)}{s(1+s)(1+3s)}$$

Improvements to the closed loop system can be made by adding a derivative action control term or velocity feedback, both in conjunction with the proportional control term.
   (i) If control is kept to just $K(s) = 1$ what are the gain and phase margins for the closed loop system?
   (ii) The controller is augmented with a derivative action term, say

$$K(s) = k_p(1+T_d s)$$

If $k_p = 1$ and $T_d = 2$ what are the new gain and phase margins?
   (iii) A velocity feedback loop from the plant is used instead of the derivative term in the series controller. Compare the use of the Bode plot in assessing this type of control addition with the addition made in (ii). If this has gain $k_v = 1$ outline the new system Bode plot and obtain the corresponding stability margins.

Answer (i) 1.33 (2.5 dB), 7.3°; (ii) ∞, 47.1°;
(iii) 2.67 (8.5 dB), 38.9°

12 A closed loop system has negative feedback with feedback transfer function $h(s)$ and forward path transfer function $g(s)$, given respectively by

$$h(s) = \dfrac{1}{1+0.1s} \quad \text{and} \quad g(s) = \dfrac{10}{(1+s)(1+0.5s)}$$

Using the inverse polar plot, (i) determine the gain and phase margins of the closed loop system, and (ii) plot $g(j\omega)^{-1}$ and combine this with

$h(j\omega)$ as a way of displaying the *closed loop* frequency response of the system.

Answer (i) 2.0 (6.0 dB), 18.8°

13 By superposition of the open loop polar plot $g(j\omega)$ on the constant M and N circles, plot the approximate closed loop frequency response phase and magnitude values, over the available frequency range, if

$$g(s) = \frac{3}{s(1+s)}$$

Check the result by evaluation of the closed loop values at chosen individual frequencies.

14 Plot the frequency response of the open loop transfer function

$$G(s) = \frac{8}{(1+0.2s)(1+0.4s)(1+0.8s)}$$

on the Nichols chart. Express $G(j\omega)$ as its magnitude and argument and evaluate sufficient specific values to define the plot clearly on the Nichols chart.

(i) What are the closed loop gain and phase margins?

(ii) At what approximate frequency is the closed loop magnification a maximum?

Answer (i) 1.41 (3.0 dB), 11.2°; (ii) 4.1 rad s^{-1}

10

Phase compensators and other controllers

The simple feedback controller may be improved by the addition of terms in series in the forward path, by using knowledge of the system to add 'feedforward' control terms, or by the addition of extra loops apart from the major feedback loop. These phase compensators, feedforward controllers and cascade control loops are important additions to the methods of improving overall system behaviour. The addition of simple derivative and integral control, velocity feedback and the use of feedforward and cascade control have been covered in earlier chapters. The use of 'phase compensators' is treated here, using the time domain (transient) response and frequency response material from the earlier chapters.

Phase compensators

Whereas the overriding consideration for control systems is stability it is also important to be able to satisfy other demands required of specific systems such as bandwidth and specified gain and phase margins. In terms of the frequency response this is reflected in the shape of the Nyquist or Bode plots which show the gain—phase characteristics of the system. The reshaping of these plots by the addition of controller terms is referred to as 'compensation' or as the addition of 'phase compensators', as these additional terms are classified according to whether they add to or decrease the open loop phase angles. Although of basic simple form individually they may be combined in attempting to shape the frequency response curves over a wide frequency range. They exhibit more flexibility, therefore, than the established PID controllers which have been the mainstay of the process industries for many years. However, similarities between the characteristics of the PID controllers and simple phase compensators are readily seen.

In broad terms the low frequency gain is an indication of steady state or slow change tracking performance, the high frequency part of the plots showing bandwidth and high frequency disturbance—rejection properties. The phase margin is a measure of closed loop stability and damping. The ability to influence these features by the use of compensators is thus significant in determining the overall performance. The basic compensator has a simple pole—zero transfer function, e.g.

$$K\,\frac{1+Ts}{1+\alpha Ts}, \quad \alpha\,\frac{1+Ts}{1+\alpha Ts}, \quad \text{or} \quad \frac{s+a}{s+b}$$

For α less than unity ($b>a$), the compensator is phase lead and for α less than unity ($b<a$), it is phase lag. A series combination of these leads to a lag–lead compensator.

The characteristics of the compensator itself are readily shown on the Nyquist or Bode plots. As the key variables α and T are varied so the gains and phases of the zero and pole factors are changed until the net desired effect on the full loop plot is achieved as far as possible. Additional phase-free gain adjustment is possible, e.g. through the gain K.

Fundamental to more detailed compensator design are Bode's theorems. These relate the phase shift with frequency to the slope of the log-magnitude against frequency plot. In particular, for non-minimum phase systems (no right half plane poles or zeros), a continuous slope on the Bode plot of ±20 dB per decade gives a phase shift of $\pm90°$, ±40 dB per decade gives $\pm180°$ and so on. Adjacent regions of the log-magnitude plot near to a specific frequency of interest, e.g. near to the gain crossover frequency, also contribute to the phase shift according to their slope. Thus to achieve a reasonably valued positive phase margin, i.e. to have a phase angle safely above $-180°$, a log-magnitude slope of the order of -20 dB per decade is required in the gain crossover frequency region.

10.1 Basic compensators

By use of the simple Bode plot construction show the gain and phase effects of the three compensators

lag, $\qquad g_c(s) = \dfrac{1+s}{1+4s}$

lead, $\qquad g_c(s) = \dfrac{1+s}{1+0.5s}$

lag–lead $\qquad g_c(s) = \left(\dfrac{1+4s}{1+40s}\right)\left(\dfrac{1+s}{1+0.1s}\right)$

Solution Consideration of the compensator Bode plots alone illustrates more clearly the contribution in shaping the overall Bode plot which will be made by each type of compensator.

(i) For the phase lag compensator the corner frequencies are at $\omega = 1$ (rad s^{-1}) for the zero and $\omega = 0.25$ for the pole. The magnitude falls from unity (0 dB) at low frequency to 1/4 (-12 dB) at the higher frequencies. The phase angle is given by

$$\arg(G(j\omega)) = \tan^{-1}\omega - \tan^{-1}4\omega$$

and thus ranges from zero at low frequencies down to a maximum lag between the corner frequencies before returning to zero at high frequency values, Fig. 10.1.

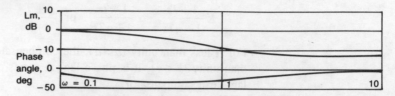

Figure 10.1 Phase lag compensator, Bode plot

Note that there is no loss of low frequency gain but that high frequency attenuation remains. The phase lag is restricted to being most significant between the corner frequencies. It is possible with this type of compensator to restrict its phase lag in the gain crossover frequency range of a system while introducing improved gain and phase margins from the attenuation introduced.

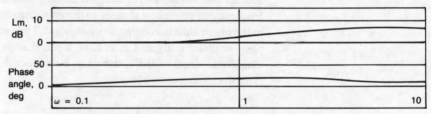

Figure 10.2 Phase lead compensator, Bode plot

(ii) For the lead compensator the corner frequencies are at $\omega = 1$ for the zero and $\omega = 2$ for the pole. The phase angle introduced is

$$\arg(G(j\omega)) = \tan^{-1}\omega - \tan^{-1}0.5\omega$$

and is thus positive for all frequencies, Fig. 10.2. It ranges from zero at low frequencies through a positive maximum between the corner frequencies to return to zero at the high frequencies. Comparison with the lag compensator

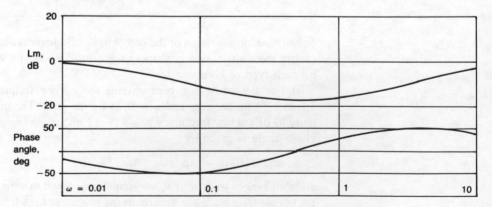

Figure 10.3 Lag−lead compensator, Bode plot

shows the magnitude now to be greater at the higher frequencies tending to 2 (6 dB). Benefit comes from the increased phase angle being placed at the region of the system gain crossover frequency in order to increase the phase margin directly. There will be some offset of this gain in phase margin due to the effect of the increase in the magnitude term.

(iii) The separation of the poles and zeros is such as still to give a benefit from the phase advance (at a system phase crossover frequency). Low frequency gain is not subject to attenuation and choice of the time constants has also kept the high frequency magnitude at unity for the compensator, Fig. 10.3. The plot illustrates how composite controllers, perhaps of higher order than this, can shape the contribution to both gain and phase over a given frequency range. Note that magnitude and phase cannot be independently adjusted over any frequency range.

10.2 Phase lead compensator and Bode plot

Although stable at all gains the second order plant represented by the block diagram in Fig. 10.4 has a low gain margin and might therefore become easily susceptible to instability arising from small unmodelled lags or delays in the system. It is not acceptable just to reduce the gain K_p as this would reduce the bandwidth below requirements. Design a simple phase lead compensator to raise the phase margin to about 50°.

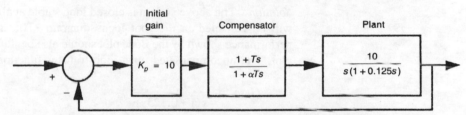

Figure 10.4 Addition of phase compensator

Solution The Bode plot for the uncompensated system, Fig. 10.5, shows the low phase margin of 15° (180° − 165°) but closed loop stability.

The maximum phase advance of about 50° is obtained from selecting a phase

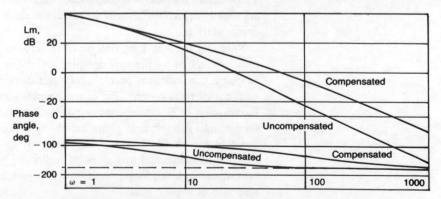

Figure 10.5 Addition of phase lead compensator

advance compensator with $\alpha = 0.1$. By choosing the compensator breakpoints (corner frequencies) to straddle the gain crossover frequency this maximum phase advance will be at or near the crossover frequency. With this crossover frequency at approximately 30 rad s^{-1} a suitable value of $1/T$ will be $30 \times \sqrt{0.1}$, say about 10 rad s^{-1}. The compensator corner frequencies are then at 10 and 100 rad s^{-1}. The effect of this compensator is shown in the second plot of Fig. 10.5. The steady state error will be zero because of the pole at the origin so there is no requirement to introduce additional gain on this count. The new phase margin is about 55° ($180° - 125°$) and further detailed tuning of α and T could be used to effect further small improvements. Note though that scope to apply simple first order compensators may be limited by closer specification of closed loop performance, such as bandwidth. In this example there is a significant increase in bandwidth on the addition of the compensator.

{It may be noted that the maximum phase angle from a compensator $(1+Ts)/(1+\alpha Ts)$ is $\sin^{-1}[(1-\alpha)/(1+\alpha)]$.}

10.3 Lead compensator and root locus

Show the result of adding the compensator in Example 10.2 to the uncompensated system by means of a sketch of the root locus diagram and hence discuss the effect on the transient response.

Solution   The above system is closed loop stable at all values of gain and this will be reflected in the root locus diagram. The improvement in system performance shown by the Bode plot should also be illustrated by the root locus plot. The uncompensated open loop transfer function is

$$G(s) = \frac{K_p\, 10}{s(1+0.125s)}$$

and with the chosen phase advance compensator becomes

$$G(s) = \frac{K_p\, 10(1+0.1s)}{s(1+0.125s)(1+0.01s)}$$

The root loci for both of these functions are shown in Fig. 10.6. These sketches are rapidly drawn sufficiently accurately to show the major effects of the added compensation.

Note that the low gain margin of the uncompensated system on the Bode plot is matched by the low damping exhibited by the branches of the roots having a constant real part of -4. The damping decreases with increasing gain but there is no crossing of the branches into the right half plane. With the addition of the compensator pole and zero the complex roots are drawn further into the left half plane indicating a greater measure of stability and more highly damped poles. The order of the system is increased by the compensator pole and this gives an additional branch to the locus, leading in turn to a real closed loop pole, fast at low gain but being slower in the overall response at higher gains.

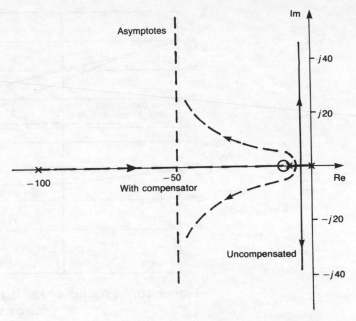

Figure 10.6 Effect of the same added compensator on the root locus plot

10.4 Phase lag compensator and Bode plot

Taking the system with open loop transfer function

$$G(s) = \frac{K}{s(1+0.025s)(1+0.05s)}, K = 75$$

a phase lag compensator is introduced to give additional damping (reduced gain) at the higher frequencies, with its own transfer function being

$$g_c(s) = \frac{1+0.25s}{1+2s}$$

Using simple Bode plots show the changes in the closed loop gain and phase margins.

Solution The Bode magnitude plot is readily drawn by addition of the independent elements for 75, $1/s$, $1/(1+0.025s)$ and $1/(1+0.05s)$ using the asymptotic construction or by full computation, Fig. 10.7. Because of the $1/s$ factor the phase angle for the uncompensated plant runs from $-90°$ to $-270°$. The magnitude plot shows a gain crossover frequency of 33 rad s^{-1} and a phase margin at this frequency of $-7°$ ($180° - 187°$), i.e. the system is closed loop unstable. (The gain margin is -2 **dB**.) See *Note* below.

The compensator has corner frequencies at 0.5 rad s^{-1} for the pole and 4 rad s^{-1} for the zero, both well below the original crossover frequency. With the addition of this phase lag compensator there is increased phase lag

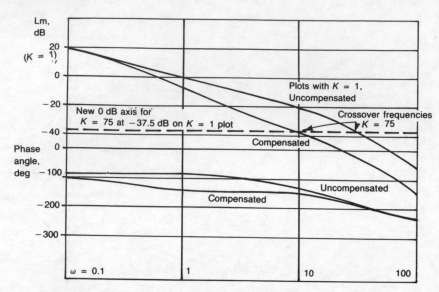

Figure 10.7 Effect of phase lag compensator on stability margins

at all frequencies but the magnitude is also attenuated. The result is a gain crossover frequency of 10 rad s^{-1} and a new phase margin of **30°** $(180° - 150°)$, despite the increased phase lag. It can be seen from the plot that the additional attenuation of $0.25/2$ at the higher frequencies, a reduction in magnitude of 18 dB, has a more significant effect than the residual phase lag from the compensator at these frequencies. The corresponding gain margin is approximately **13 dB**, determined from the plot.

Note. The effect of a constant gain on the Bode plot is to move the gain curve vertically as a whole. For example, a gain of 20 is equal to 26.0 dB and the plot is lifted by this amount. A gain of 0.5 is a shift of -6.0 dB and so on. Alternatively the magnitude plot for a fixed gain of unity may be plotted and the effect of changing the constant gain factor K is then shown by moving the frequency axis up or down by the same amount but in the opposite sense. Thus here the gain of $K = 75$ has been shown by drawing the magnitude plot for $K = 1$ and then considering the intersection of the magnitude plots with the -37.5 dB line ($-20 \log 75$).

10.5 Lag compensator and root locus

> Show the effect of the above compensator on the root locus plot. How is the transient response affected?

Solution The root locus for the uncompensated system has three branches corresponding to the three real open loop poles and takes up the familiar shape as shown in Fig. 10.8. It has been constructed by evaluation of the asymptotes, their intersection, the breakpoint on the real axis and the intersection with the imaginary axis. The critical gain of 60 for stability, compared with the system gain of 75, shows that the given system is closed loop unstable with the gain

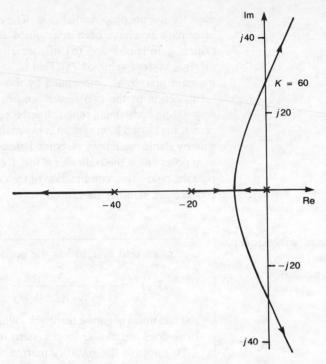

Figure 10.8 Root locus of uncompensated system

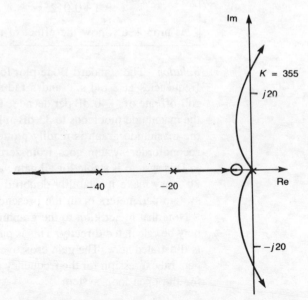

Figure 10.9 Root locus with compensator added

being a factor of 75/60 too great. This corresponds to 2 dB, agreeing with the result from the above Bode plot.

With the addition of the pole and zero from the compensator the plot is more complex but may still be plotted, Fig. 10.9, using the same key elements as used for the uncompensated case. The critical gain and crossing point on the imaginary axis have been determined using the Routh stability criterion. The critical gain is now 355 (51 dB) which is a factor of 355/75 greater than the original system value of 75. That is, it is 13.5 dB greater, again confirming the gain margin as determined by the Bode plot.

Inspection of the two figures shows that in both cases there is a rapidly responding component in the closed loop system response. For a range of low gain K the closed loop system is overdamped. This is followed with increasing gain by stable oscillatory response before eventually instability occurs as closed loop poles fall in the right half plane. Despite the introduction of the additional pole the pole−zero combination of the compensator brings additional damping and greater stability margins.

10.6 Phase advance compensator to improve phase margin

A motorized system has the open loop transfer function

$$G(s) = \frac{40}{(1+0.2s)(1+0.5s)}$$

and has unity negative feedback. What is the closed loop phase margin? How does the closed loop system respond to a unit step input?

To improve the stability margin a phase advance compensator

$$g_a(s) = \frac{1+0.1s}{1+0.0125s}$$

is proposed. Show the effect of this by using a Bode plot.

Solution The standard Bode plot for the uncompensated system has corner frequencies at 2 rad s^{-1} and 5 rad s^{-1} and tends to a final high frequency roll-off rate of -40 dB per decade. The steady state open loop gain is 40, i.e. the magnitude plot tends to 32 dB at low frequencies. Based on these features the magnitude graph is readily plotted, Fig. 10.10. The phase angle for the second order system goes from zero to $-180°$ at the higher frequencies.

This system is stable for all gain values but a lowish phase margin of about 20° may make its stability doubtful in the presence of small changes in the system parameters or in the presence of unmodelled dynamics.

Note that in addition to the graphical methods the gain and phase margins may be calculated directly. This is particularly so with low order systems and is illustrated here. The gain crossover frequency is determined by setting the general expression for the frequency response magnitude equal to unity. Then for the open loop system

$$G(s) = \frac{40}{(1+0.2s)(1+0.5s)}$$

the gain is

$$|G(j\omega)| = \frac{40}{\sqrt{(1+0.04\omega^2)}\,\sqrt{(1+0.25\omega^2)}}$$

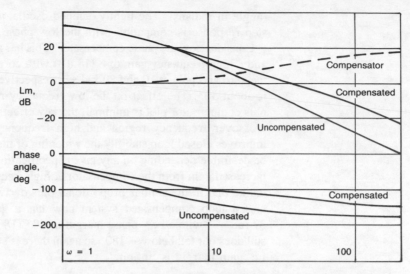

Figure 10.10 Addition of phase lead compensator to second order system

and equating this magnitude to unity gives

$$0.01(\omega^2)^2 + 0.29\omega^2 - 1599 = 0$$

yielding the positive root $\omega^2 = 385.5$, i.e. $\omega = 19.6$.

The phase margin may now be calculated as at this frequency the phase angle is

$$\arg(G(j\omega)) = -\tan^{-1}(0.2 \times 19.6) - \tan^{-1}(0.5 \times 19.6)$$
$$= 159.9°$$

to give a phase margin of **20.1°** in agreement with the Bode plot derived figure. As the phase angle never falls below $-180°$ there is no finite corresponding gain margin.

For the closed loop step response of this system the closed loop transfer function is seen to be

$$\text{CLTF} = \frac{40}{0.1s^2+0.7s+41}$$

$$= \frac{400}{s^2+7s+410}$$

The steady state gain is 40/41, the undamped natural frequency $\omega_n = \sqrt{410}$ ($= 20.25$ rad s^{-1}) and the damping coefficient $c = 7/(2\sqrt{410}) = 0.173$. The standard response form to a unit step input for a second order system gives for the output, say x,

$$x(t) = \frac{40}{41}[1-1.015e^{-3.5t}\sin(19.94t+1.397)]$$

(angle in radians). The lightly damped oscillatory, but stable, nature of the step response is compatible with the low phase margin calculated.

Consider now the phase compensator. This has a low frequency gain of unity and a high frequency gain of 8 (18 dB) with corner Bode plot frequencies of the zero and pole at 10 and 80 rad s^{-1} respectively. The peak phase advance is about 50°. The effect on the low frequency range of the combined plant plus compensator plot is minimal, the key effects being phase advance in the crossover frequency region and high frequency amplification. While this improves closed loop stability the widening of the bandwidth may or may not be desirable depending on any given specification. Note that the (open loop) increased gain from the compensator at high frequencies is, however, limited unlike that in the similar proportional plus derivative action controller. The closed loop compensated system now has a gain crossover frequency of 37 rad s^{-1} with a new phase margin of **60°** (180° − 120°). The phase angle still does not fall below − 180° so again there is stability for all values of gain, i.e. gain margin is 'infinite'.

10.7 A phase lag compensator with the same system

It is suggested that a phase lag compensator

$$g_I(s) = \frac{1+0.625s}{1+5s}$$

might be used instead for the above system, Example 10.6. Show the effects of this also on the Bode plot, commenting on the relative changes in the open loop dynamic gains and phases.

Discuss the relative merits of these two schemes in terms of the closed loop system transient response and the ability to meet a possible requirement for increased steady state gain.

Solution When using the phase advance compensator above it was seen that there was little effect on the low frequency end of the Bode plot. The phase lag compensator now considered has larger time constants and hence gives corner frequencies at the lower end of the frequency range. In this case the pole−zero pair have corner frequencies at 0.2 and 1.6 rad s^{-1} respectively. The new phase angle from these is negative but peaks below the crossover region of the uncompensated system. The magnitude of the compensator is less than unity at all frequencies, in particular tending to a factor of 0.625/5 at the higher frequencies, an attenuation of − 18 dB. These features are shown in the plots of Fig. 10.11, which are now shown extended over an additional lower decade of frequency.

From the original phase margin of 20° at the gain crossover frequency of 20 rad s^{-1} the phase margin is improved to **43°** (180° − 137°), at a new crossover frequency of 6.3 rad s^{-1}. The bandwidth has been considerably reduced and this may be disadvantageous if rapid accurate tracking in a transient response is required. Some adjustment of gain might be acceptable to reduce this fall in bandwidth and at the same time improve the steady state gain, thereby reducing steady state error, while still retaining some of the improvement in

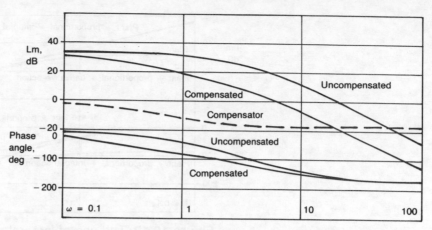

Figure 10.11 Addition of phase lag compensator to the same second order system

the stability determining phase margin. For example, increasing the gain by a constant factor of 2 (6 dB) increases the gain crossover frequency to 9 rad s^{-1}, the steady state gain from 40 to 80, i.e. from 32 to 38 dB, and reduces the steady state error by approximately half from 2.4% (1/41) to 1.2% (1/81). The new phase margin would be about 33° (180° − 147°), still an improvement on the uncompensated system value.

These two examples show the differences between the phase advance and phase lag compensators when used on the same simple system. The choice is influenced not just by the need to improve the phase margin but also by the requirements in terms of transient response such as bandwidth, speed of tracking and steady state errors. The associated features are shown also by the root locus plot as indicated by the earlier examples.

The relationship between these phase compensators and the pure addition of derivative and integral action modes is illustrated further by the following example.

10.8 Derivative and integral action with the same system

The lead and lag compensators above may be seen as realization approximations to the addition of derivative and integral control modes respectively and having the advantage of limiting the amplification of high frequency inputs (noise) which occurs with derivative action. For the previous system of Example 10.6 establish the Bode plot for the introduction of the PD and PI controllers,

$$g_1(s) = K(1+0.1s), \quad g_2(s) = K\left(1 + \frac{1}{0.625s}\right)$$

with K initially set to unity. Compare the results with the Bode plots for the previous examples for phase advance and phase lag compensators and again check the closed loop system dependence on the gain K.

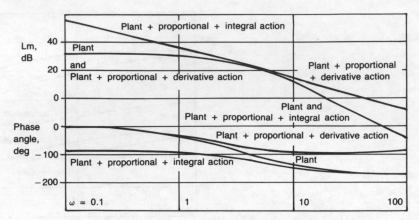

Figure 10.12 The use of integral and derivative action
in place of the lag and lead compensators

Solution (i) Consider the derivative action first. The magnitude of this PD term and the phase angle for the Bode plot contributions are

$$|g_1(j\omega)| = \sqrt{(1+0.01\omega^2)}, \quad \arg(g_1) = \tan^{-1}0.1\omega$$

giving a single corner frequency at 10 rad s^{-1} and magnitude increasing at 20 dB per decade at higher frequencies. The phase angle increases from zero at low frequencies to 90° at higher frequencies. The effect of this is that the low frequency section of the plant and plant plus controller are very similar but at high values the roll-off rate is reduced to −20 dB per decade, Fig. 10.12, compared with −40 dB per decade when using the phase advance compensator. In the gain crossover region the plots for PD action and phase advance compensation are similar also but with there being greater net phase advance from the controller in the former case. The phase margin in this case is **86°** (180° − 94°).

(ii) Discussion of the PI controller is similar. Although only a single additional control mode is added, the effect is to establish both a new pole and zero,

$$g_2(s) = K\left(1 + \frac{1}{0.625s}\right) = \frac{K}{0.625}\left(\frac{1+0.625s}{s}\right)$$

At low frequency the pole (s) dominates giving a magnitude changing at −20 dB per decade and passing through the (0 dB, 1 rad s^{-1}) point. At high frequencies the controller tends to K and with $K = 1$ the Bode plots for the plant and plant plus PI control become the same. Thus there is no benefit in high frequency tracking or bandwidth. However, the low frequency magnitude shows that steady state error tends to zero. Maximum phase lag from the controller is 90° at low frequency and zero at the high frequency range. In the gain crossover region the phase margin is in fact reduced to **15°** (180° − 165°). Comparison with the phase lag compensator plot shows that this is because of the absence of reduced gain with integral action rather than difference in phase lag. However, the gain K could now be reduced to improve the phase margin but maintain the zero steady state error, and the integral action time (0.625) could be increased to reduce the coefficient of $1/s$ in the controller.

These results illustrate the similarities, and also the differences, between phase advance compensation and derivative action and phase lag compensation and integral action.

10.9 Use of the Bode plot with the PID controller

Note that the product of compensator terms in the above problem, $g_1(s)g_2(s)$ with $K = 1$, gives a 'PID' controller

$$g_3(s) = 1.16 + 0.1s + \frac{1}{0.625s}$$

Use this compensator to illustrate the combined effects of the terms in this controller.

Solution It was seen in the previous example that the derivative and integral actions essentially reshaped in different ways and over different ranges of frequency the Bode plot for an initial system. Hence it might be anticipated that the combined use of these additional terms, or the combination of both phase advance and phase lag compensators as a lag−lead compensator, could give rise to the benefits of both control modes. Thus although integral action removes steady state offset it leads alone to a reduction in phase margin while derivative action primarily improves the phase margin.

The proportional plus integral plus derivative action (PID) controller used in this question is a product of the PI and PD controllers used above, the gain factor K in each case being set to unity. Its magnitude and phase contributions are best evaluated with this in mind rather than proceeding by way of using the $g_3(s)$ expression as given. Thus the magnitude term is

$$|g_3(j\omega)| = |g_1(j\omega)| \cdot |g_2(j\omega)|$$

$$= \sqrt{(1+0.01\omega^2)} \cdot \left(\frac{\sqrt{(1+0.625^2\omega^2)}}{0.625\omega} \right)$$

and the phase angle

$$\arg(g_3(j\omega)) = \arg(g_1(j\omega)) + \arg(g_2(j\omega))$$

$$= \tan^{-1}0.1\omega + \tan^{-1}0.625\omega - 90°$$

The Bode plot for the PID controller is shown separately in Fig. 10.13. The magnitude plot shows the characteristic of the integral action with its initial slope of -20 dB per decade and high gain at low frequency, coupled at the high frequency range with a slope of $+20$ dB per decade. Note that the independent choice of the integral and derivative parameters gives flexibility in shaping the controller Bode plot and hence in shaping that of the controlled system. The phase shift is similarly dominated at low frequency by the integral term and at higher frequencies by the derivative term. Note that the magnitude and phase plots cannot be manipulated independently of each other. This PID controller may be considered as a specific example of a lag−lead compensator.

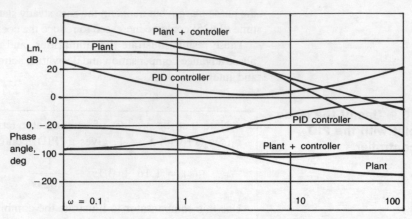

Figure 10.13 Use of the PID controller with the Bode plot

[The full (plant plus controller) magnitude plot has zero corner frequencies at 1.6 and 10 rad s^{-1} and pole corner frequencies at 2 and 5 rad s^{-1}.]

The effect of this controller on the second order plant may now be seen. The Bode plot shows the low frequency, steady state, benefit of the integral action. At the same time the derivative term keeps the bandwidth up and more than counters the phase disadvantage of the integral term. The overall system thus has good steady state performance, maintained tracking ability and a high phase margin of about 82° (180° − 98°). The PID controller has thus retained the separate advantages of the PD and PI controllers. The further shaping of a Bode plot to establish a system which gives a specified closed loop performance is given in the next example.

10.10 Fuller compensator design

Full design of a single-input single-output controller may require the meeting of specifications of transient behaviour and noise rejection properties. These may be conflicting, some compromise with a simple controller may be required, and the solution may not be unique but one of a range of limited possibilities. This example is intended to illustrate such a situation, which could be aided by a computer based control design package, using the shaping of the Bode plot as the basis of the design.

Figure 10.14 represents a simple feedback control system which is, however, subject to disturbances, shown as input $d(s)$ on the plant output (within the loop) and as noise $n(s)$ on the feedback (measurement) path. It is required that the closed loop achieves a phase margin of not less than 40° while satisfying the additional requirements that the output disturbance at frequencies less than 2 rad s^{-1} and the measurement noise at frequencies greater than 100 rad s^{-1} are reduced to less than 5% of their source values. Design a simple compensator $K(s)$ to match these requirements.

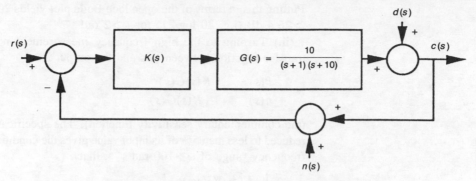

Figure 10.14 System with output disturbance and measurement noise

Solution This example widens the use of compensators within a more detailed control system design so that a performance specification including noise and disturbance rejection may be met. Complex demands may sometimes require conflicting requirements from simple compensators and much tuning and additional terms will be added by the use of computer aided control design packages. However, some basic steps are indicated by this example.

(i) Consider the uncompensated system's Bode plot. The magnitude is unity (0 dB) at low frequency and with corner frequencies at 10 and 1 rad s^{-1} it has a high frequency roll-off rate of -40 dB per decade. The phase angle runs from zero to $-180°$. The Bode plot is shown in Fig. 10.15.

(ii) To express the closed loop specification in terms of the open loop Bode plot it is necessary to look at the closed loop to open loop relationships. First consider the requirement to suppress the low frequency (<2 rad s^{-1}) disturbance $d(s)$. The closed loop transfer function between this disturbance and the output $c(s)$ is the sensitivity function

$$\frac{c(s)}{d(s)} = \frac{1}{1 + K(s)G(s)}$$

The requirement that the magnitude of $d(s)$ is reduced at a frequency ω to be only 5% at the output $c(s)$ of what it is at its source (or if the system is open loop) is that the magnitude of this transfer function throughout the nominated frequency range is less than 0.05, i.e.

$$\left| \frac{1}{1 + K(j\omega)G(j\omega)} \right| < 0.05$$

or

$$|1 + K(j\omega)G(j\omega)| > 20 \dots\dots\dots\dots\dots\dots\dots\dots\dots\dots\dots\dots\dots\dots\dots \text{[i]}$$

Now for $|K(j\omega)G(j\omega)| > 1$, as it must be here to achieve the attenuation, the general condition $|1 + K(j\omega)G(j\omega)| > |K(j\omega)G(j\omega)| - 1$ holds. Then the above inequality [i] must be satisfied also if $|K(j\omega)G(j\omega)| - 1 > 20$. It is sufficient then that $|K(j\omega)G(j\omega)| > 21$ and this condition will, if anything, be on the conservative side giving a $K(s)$ which should readily satisfy [i].

Putting this in terms of the open loop Bode plot yields $20 \log |K(j\omega)G(j\omega)|$ >26.4 dB (i.e. $20 \log 21$) for $\omega > 2$ rad s^{-1}.

(iii) Turning to the high frequency measurement noise the closed loop transfer function between this and the output is

$$\frac{c(s)}{n(s)} = \frac{K(s)G(s)}{1+K(s)G(s)}$$

(the complementary sensitivity function). The specification that the noise is reduced to less than 5% of its input value gives the condition, over the specified frequency range of $\omega > 100$ rad s^{-1}, that

$$\left| \frac{K(j\omega)G(j\omega)}{1+K(j\omega)G(j\omega)} \right| < 0.05$$

or

$$|1+K(j\omega)G(j\omega)| > 20|K(j\omega)G(j\omega)| \quad \dots\dots\dots\dots\dots\dots\dots \text{[ii]}$$

By a similar general relationship to that used above, if $|K(j\omega)G(j\omega)| < 1$, as is necessary for this noise reduction, then $|1+K(j\omega)G(j\omega)| > 1 - |K(j\omega)G(j\omega)|$. Using this condition with equation [ii] gives the sufficient condition $1 - |K(j\omega)G(j\omega)| > 20|K(j\omega)G(j\omega)|$ or $|K(j\omega)G(j\omega)| < 1/21$ for $\omega > 100$ rad s^{-1}. Expressed suitably for the Bode plot this is $20 \log |K(j\omega)G(j\omega)| < -26.4$ dB.

(iv) It is necessary to reconcile both of these conditions with the requirement on phase margin of not less than 40°. The conditions of paragraphs (ii) and (iii) are shown on the Bode plot by the hatched regions A and B in Fig. 10.15.

It is necessary that the Bode plot as modified by any compensator does not pass through these hatched regions. Inspection of the Bode plot shows that simply increasing the system open loop gain by a constant 33.5 dB ($K = 47$) might just result in the magnitude plot passing between these restrictions. However, such an increase would give a gain crossover frequency of 20 rad s^{-1} and a phase margin of just under 30°, less than required. To

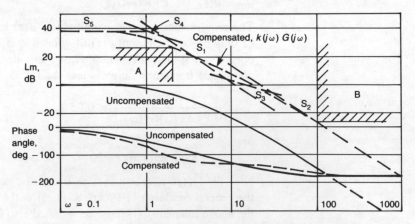

Figure 10.15 Design of compensator using the Bode plot

achieve a phase margin of at least 40° an increase in constant gain of only 28 dB ($K = 25$) can be allowed. At this value of gain the reduction on the output disturbance $d(s)$ would not be met. To meet all the specification a more complex compensator is needed.

To improve the phase margin a phase advance compensator is a first consideration. However, this will reduce the attenuation at high frequencies while doing little with the low frequency gain. Additional constant gain gives the difficulty as before. Progress may be made by looking not so much at the compensator and plant Bode plots separately but at the final shape of plot which is required from their combination. Recall that the magnitude plot may be approximated by straight line asymptotic sections having discrete slope values of ± 20 dB per decade, ± 40 dB per decade, etc. The corresponding phase angles are $\pm 90°$, $\pm 180°$, etc., but they are also influenced by the rest of the magnitude slopes, especially those adjacent to the section under consideration. (The full conditions are given by Bode's laws.)

The specification means that between the frequencies of 2 rad s^{-1} and 100 rad s^{-1} the magnitude plot must fall in excess of 50 dB. As this occurs over less than two decades it can only be achieved if there are sections with a slope in excess of -20 dB per decade, i.e. of -40 dB per decade or more. Because of the magnitude slope to phase angle relationship we need a slope of -20 dB per decade in the region of gain crossover, this slope being associated with the lower phase lag of 90° and hence a better phase margin. Hence selection of a magnitude plot with a central slope of -20 dB per decade and outer regions (but between the specified frequencies of 2 and 100 rad s^{-1}) of -40 dB per decade would appear desirable. The final asymptotic magnitude plot then appears as in Fig. 10.15, where at low frequencies the shape of the plot is determined by the plant and compensator constant gain factor, in the mid range by this gain plus the plant and the dynamic terms of the compensator, and in the high frequency range by the plant and gain again.

The asymptotes S_1 and S_2, each of slope -40 dB per decade, are drawn first. To allow for the true magnitude in the region of the corner frequencies S_1 is drawn to exceed the critical value at $\omega = 2$ rad s^{-1} by about 4 dB. As the true plot falls below the asymptote this is not necessary at $\omega = 100$ rad s^{-1} where the line is drawn to pass just outside the prohibited region. An asymptote S_3 of slope -20 dB per decade is drawn in the mid section cutting the 0 dB axis about midway between the lines S_1 and S_2. This minimizes at the gain crossover frequency the higher phase lag contributions from the sections of higher slopes. Note that there is some flexibility in the positioning of the asymptotes and in the corresponding corner frequencies which subsequently occur. The solution is not therefore unique. For example, the positioning of the section S_4 is open to variation. If the corner frequency here is kept to that of the original plant then the movement of S_4 in turn determines the exact positioning of S_5, the precise low frequency gain, and the corner frequency of S_4 with S_1. However, this flexibility of choice is still quite restricted and more detailed or tighter closed loop specification could require additional sections of plot, i.e. a more complicated compensator.

The corner frequencies from the asymptotic plot give the compensated system poles and zeros. There are corner frequencies at 1, 1.5, 8.4, and 32 rad s^{-1}.

Noting the changes in slope at each of these gives us open loop poles at 1, 1.5, and 32 rad s^{-1} and a zero at 8.4 rad s^{-1}. The combined compensated system has the **open loop transfer function**

$$K(s)G(s) = \frac{K'(s+8.4)}{(s+1)(s+1.5)(s+32)}$$

The gain crossover frequency is 15 rad s^{-1}. Putting $|K(j\omega(G(j\omega)| = 1$ at $\omega = 15$ enables the value of K' to be determined from

$$1 = K' \sqrt{\left[\frac{(\omega^2+8.4^2)}{(\omega^2+1)(\omega^2+1.5^2)(\omega^2+32^2)}\right]}$$

so that $K' = 465$. The open loop magnitude can now be checked at $\omega = 2$ and 100 rad s^{-1} to give 22.4 (= 27.0 dB) and 0.044 (= -27 dB) respectively. The phase margin is calculated from the phase angle at $\omega = 15$ rad s^{-1}. The phase angle is

$$\tan^{-1}\left(\frac{15}{8.4}\right) - \tan^{-1}\left(\frac{15}{1}\right) - \tan^{-1}\left(\frac{15}{1.5}\right) - \tan^{-1}\left(\frac{15}{32}\right)$$

i.e. $-135°$ to give a phase margin of $45°$, an acceptable value.

The compensator $K(s)$ is obtained by taking out the factors of $G(s)$, i.e. $10/(s+1)(s+10)$ to leave

$$K(s) = \frac{46.5(s+8.4)(s+10)}{(s+1.5)(s+32)}$$

containing both lead and lag elements.

Feedforward control

Feedforward control has been considered in Chapter 7. It constitutes an addition to the basic feedback controller involving the generation of more complex pole/zero terms and has this in common with the more extended compensators.

Cascade control

This has also been treated in Chapter 7 and again represents an extension upon the basic controller.

Loop interaction

When more than one controller is used on a more complex system the principles of single-input single-output control may still be applied although now there is likely to be difficulty caused by interaction between the loops. If this interaction is severe and if there are a number of variables to be considered then the more recent principles of multi-variable controllers are used to advantage. However, the effect of such in-plant interaction is shown by using simplified systems which nevertheless illustrate the potential difficulties. It may not be possible to remove such effects with simple controllers and more fundamental changes to the system itself may be beneficial.

10.11 Derivation of transfer function in system with interaction

Interactions of control loops are common within process systems and simple representation is usually difficult. For the block diagram representation shown in Fig. 10.16 determine the transfer function relating the effect of the demanded change r_1 in c_1 on the second output variable c_2.

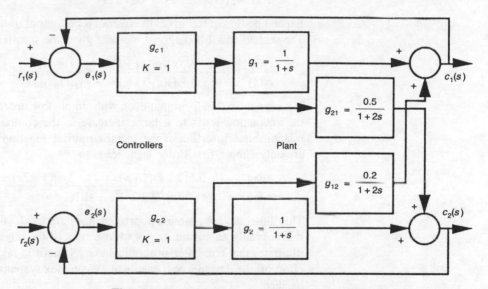

Figure 10.16 System with process interaction

Solution From the block diagram the individual internal relationships between variables are set up and combined into an overall input/output vector—matrix form. From this the separate input-to-output equations are determined.

Directly from the diagram there are the following relationships:

$$c_1(s) = g_1(s)e_1(s) + g_{12}(s)e_2(s)$$
$$c_2(s) = g_2(s)e_2(s) + g_{21}(s)e_1(s)$$

which combine as

$$c(s) = G(s)\, e(s)$$

with $c = [c_1\ c_2]^T$ and $e = [e_1\ e_2]^T$ and G the matrix of transfer functions

$$G(s) = \begin{bmatrix} g_1(s) & g_{12}(s) \\ g_{21}(s) & g_2(s) \end{bmatrix}$$

Now the errors are given by

$$e_1(s) = r_1(s) - c_1(s)$$
$$e_2(s) = r_2(s) - c_2(s)$$

so that, using the notation as before,

$$e(s) = r(s) - c(s)$$

Hence

$$c(s) = G(s)[r(s) - c(s)]$$

which rearranges to give

$$c(s) = [G(s) + I]^{-1} G(s)r(s)$$

Expanding this in full gives the overall input/output transfer function matrix. The second row off-diagonal element gives the transfer function

$$\frac{c_2(s)}{r_1(s)} = \frac{-g_{21}(s)(g_1(s)+1) + g_1(s)g_2(s)}{(g_1(s)+1)(g_2(s)+1) - g_{12}(s)g_{21}(s)}$$

It is seen from this form that even with simple low order contributing factors the interaction leads to a large increase in the complexity of input/output relationships. Insertion of the specific transfer functions given, all of which are only simple first order lags, leads to

$$\frac{c_2(s)}{r_1(s)} = \frac{[-0.5(2+s)(1+s) + (1+2s)](1+2s)}{(2+s)^2(1+2s)^2 - 0.1(1+s)^2}$$

This illustrates why simple approximations are used when possible although it is normally an advantage not to lose sight of the physical problem. It also illustrates that computer programs have a benefit in taking the manipulative effort off the designer and enable more complex systems to be systematically handled.

10.12 Control to remove interaction

For the above system introduce controllers to remove the secondary coupling between inputs and outputs and comment on the problems which may arise.

Solution A concept for the removal of the interaction between specific inputs and outputs is the introduction of additional controllers specifically designed to remove the unwanted interactions and relying on a knowledge of the plant model. However, as will be demonstrated in a simple configuration, these extra controllers have the nature of a feedforward controller and for complete dynamic cancellation rely on perfect modelling and perfect implementation of complex terms. As such the implementation is restricted although there is

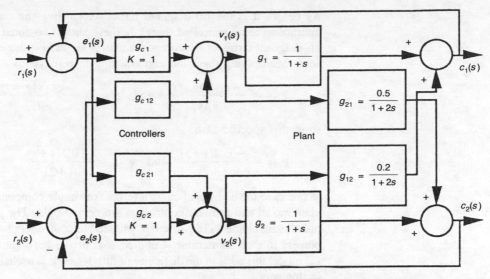

Figure 10.17 Addition of controllers to remove process interaction

potential for some steady state improvement. Similar control terms have been introduced, for example, in robotics.

The objective here is for the output c_1 to respond solely to demands r_1 and for c_2 to respond solely to demands r_2. The additional controllers are added as in Fig. 10.17.

The control input to each controlled variable of the plant, v_1 and v_2, is now the sum of two control terms,

$$v_1(s) = g_{c1}(s)e_1(s) + g_{c12}(s)e_2(s)$$
$$v_2(s) = g_{c2}(s)e_2(s) + g_{c21}(s)e_1(s)$$

or

$$v(s) = G_c(s)e(s)$$

where $v = [v_1 \ v_2]^{\mathrm{T}}$, $e = [e_1 \ e_2]^{\mathrm{T}}$ and the new controller matrix is given by

$$G_c(s) = \begin{bmatrix} g_{c1}(s) & g_{c12}(s) \\ g_{c21}(s) & g_{c2}(s) \end{bmatrix}$$

If this equation is incorporated with the previous equations from the above example, bearing in mind that the controlled inputs to the plant, v_1 and v_2, now have this general form as distinct from just being e_1 and e_2 as they were before when the only control gains were unity, then

$$c(s) = G(s)v(s)$$
$$= G(s)G_c(s)e(s)$$

and the overall input/output equation becomes

$$c(s) = [I + G(s)G_c(s)]^{-1}G(s)G_c(s)r(s)$$

As before it is the off-diagonal terms which give the interaction. For the interaction to be cancelled therefore these should be equal to zero. Now the off-diagonal terms of the composite matrix are zero if those of $G(s)G_c(s)$ are zero. Multiplying these out and equating to zero yields

$$g_{c\,12} = \frac{-g_{12}(s)g_{c2}(s)}{g_1(s)} \quad \text{and} \quad g_{c21} = \frac{-g_{21}(s)g_{c1}(s)}{g_2(s)}$$

so in this specific case

$$g_{c\,12} = \frac{-0.2(1+s)}{1+2s} \quad \text{and} \quad g_{c21} = \frac{-0.5(1+s)}{1+2s}$$

In this case the additional control terms are simple compensators because the plant model and primary controllers are very simple. The presence of higher order plant elements in particular will lead to controllers having higher order powers in their numerator. Coupled with errors between the real plant and the model this adds to the difficulty of full dynamic cancellation of interactions in this way.

Problems

Note. Once again computer based implementations are a possible route to the solution of the following problems.

1 Show the characteristics of the following compensators on the Bode and Nyquist plots, relating the two types of plot to each other by considering the gain and phase changes with frequency:

(i) $k_1(s) = \dfrac{1+0.5s}{1+5s}$ (ii) $k_2(s) = \dfrac{1+0.3s}{1+0.03s}$

What is the principal function of such compensators?

2 Plot the Bode plot for the plant/controller having the transfer function

$$g(s) = \frac{10}{(1+s)(1+0.2s)(1+0.1s)}$$

(i) What are the closed loop gain and phase margins?
(ii) A phase advance compensator with transfer function

$$k(s) = \frac{1+0.4s}{1+0.04s}$$

is added to the loop controller. What are the new gain and phase margins? What is the effect of the compensator on the system bandwidth?

Answer (i) 1.98 (5.9 dB), 21.9°, (ii) 2.72 (8.7 dB), 33.6°

3 Using the root locus diagrams for the uncompensated and compensated system in problem **2** as a basis, describe the effect of the compensator on the transient response characteristics of the closed loop system.

4 Plot the Bode plot for the open loop system transfer function

$$g(s) = \frac{K}{(1+5s)(1+20s)(1+10s)}$$

using $K = 1$.

(i) What is the maximum value of K for stability of the closed loop?

(ii) To obtain higher closed loop steady state (low frequency) gain the value of K is increased from unity to 10. What are the stability margins now? What is the effect on the stability margins of increasing K to 40?

(iii) To improve stability margins, introduce additional high frequency attenuation and reduced bandwidth, while maintaining high low frequency gain, a phase lag compensator

$$k(s) = \frac{1+25s}{1+250s}$$

is added to the proportional control ($K = 10$) of the loop. Add this compensator to the Bode plot and determine the new gain crossover frequency and the gain and phase margins.

Answer (i) 11.25, (ii) 1.125 (1.0dB), 3.7°, negative (unstable); (iii) 7.36 (17.3 dB), 68.9°, 0.04 rad s^{-1}

5 Draw a root locus diagram for the system $g(s)$ in problem **4** showing its principal features. Add the compensator $k(s)$ pole and zero and show how these modify the plot. Relate these diagrams and the changes due to the compensator to the corresponding Bode plots of problem **4**.

6 A system has the open loop transfer function

$$g(s) = \frac{4}{s(1+0.3s)(1+0.6s)}$$

Add a suitable phase lead compensator $k(s)$ to produce a phase margin in the region of 50°. Confirm that the phase margin is as desired by plotting the Bode plot for the completed open loop, $g(s)k(s)$.

7 A third order system and proportional controller have the open loop transfer function

$$g(s) = \frac{50}{(1+0.01s)(1+0.2s)(1+0.5s)}$$

(i) Plot its Bode plot and determine the gain and phase margins.

(ii) What is the limiting value of controller gain (given as $K = 10$ in $g(s)$) if closed loop stability is to be maintained?

(iii) A phase advance controller

$$k(s) = \frac{1+0.08s}{1+0.008s}$$

is added to improve the phase margin keeping the controller constant gain factor as $K = 10$. What are the new gain and phase margins?

Answer (i) 1.49 (3.5 dB), 5.9°; (ii) 14.9;
(iii) 5.05 (14.1 dB), 45.1°

8 The compensator in problem **7**, while improving the stability margin, adds an additional unwanted 20 dB of attenuation at the higher frequencies. To counteract this the compensator is modified so that it becomes a lead–lag compensator

$$k(s) = \frac{(1+0.08s)(1+0.8s)}{(1+0.008s)(1+8s)}$$

Add the effects of this to the Bode plot.
  (i) Show that the required effect is achieved at the high frequency range.
  (ii) How much effect does this addition have on the gain and phase margins?
  (iii) What would be the effect of the additional phase lag pole and zero if used as the only compensator terms?

9 Within a phase compensator there are two parameters, in addition to gain, which can be varied. With the classical 'process control' terms of derivative action (comparable with lead) and integral (comparable with lag) there is only one in each. For the plant with open loop transfer function

$$g(s) = \frac{1}{(1+s)(1+0.2s)(1+0.1s)}$$

apply the proportional plus derivative controller

$$k(s) = K(1+T_d s)$$

with $K = 10$ and derivative action time $T_d = 0.4$. Draw the Bode plot for $g(s)k(s)$ and compare the result with the phase lead compensator of problem **2**. What is the effect of increasing the amount of derivative action?

10 The above system (problem **9**) has a steady state closed loop gain of only 0.91. To remove the subsequent error integral action is added as the additional controller term, i.e.

$$k(s) = K\left(1+ \frac{1}{T_i s}\right)$$

If $K = 10$ and $T_i = 10$ draw the Bode plot for the controller and combine this with that for the plant noting the effect on the stability margins?

11 Show that although there is some interaction between the integral and derivative terms the two modes as used in problems **10.9** and **10.10** may be combined to give a PID controller expressed as

$$K'\left(1+T_d's+\frac{1}{T_i's}\right)$$

and evaluate K', T_d' and T_i'. Putting $K = 10$ show this full PID controller and its effect on the plant using the Bode plots.

Answer $K' = 9.615$, $T_d' = 0.416$, $T_i = 9.615$

12 A motor driven device has an open loop transfer function between its demand and the output position of

$$G(s) = \frac{100}{s(1+0.01s)(1+0.04s)}$$

(i) Draw the Bode plot and assess the closed loop stability.
(ii) Add a phase lead compensator

$$k(s) = \frac{1+Ts}{1+\alpha Ts}$$

selected so that the -40 dB per decade section of the plot is modified to a slope of -20 dB per decade and a phase margin of about $40°$ is reached. Confirm your design by the Bode plot of $k(s)G(s)$.

(iii) Why is the compensator chosen in this way? What is the effect on open loop low and high frequency gain and on closed loop bandwidth?

13 A phase lag compensator might also be used for the system of problem **12**. Compare such an approach in its shaping of the Bode plot and on the final closed loop behaviour.

14 A system as shown in Fig. 10.14 is subject to two unwanted sources of 'input', $d(s)$ and $n(s)$. By considering the closed loop response of the output $c(s)$ show that for the general case unrestricted suppression of both of these inputs over the same frequency range is not possible.
  The plant dynamics are described by the transfer function

$$g(s) = \frac{1}{(1+0.2s)(1+2s)}$$

The measurement noise at frequencies beyond 50 rad s^{-1} must be reduced to no more than 2% of its input value and the process noise by at least a factor of 20 below frequencies of 1 rad s^{-1}. Design a phase compensator to achieve these needs and check that it gives a phase margin of about $40°$ or more. Draw the Bode plot to confirm the acceptability of the design. (Note that there is normally no unique solution.)

15 A two-input two-output system is as described by Fig. 10.17. The elements of the loops are now given by

$$g_{c1} = 1 \qquad\qquad g_{c2} = 1$$

$$g_1 = \frac{1}{s(1+0.5s)} \qquad\qquad g_2 = \frac{1}{1+0.8s}$$

$$g_{21} = \frac{0.1}{1+3s} \qquad\qquad g_{12} = \frac{0.2}{1+4s}$$

Draw the Bode plots for each of the controllers g_{c12} and g_{c21} which could remove the plant interaction. Discuss these in terms of the practicality of achieving complete removal of the unwanted interaction effects.

Index

additional controller terms, 128
argument, 93
attenuation, 118
auxiliary equation, 157

bandwidth, 204
block diagram, 2, 8, 53
   algebra, 53
   for cascade controller, 148
   for complex system, 64
   from physical system, 60
   reduction, 60
   with secondary input, 62
Bode plot, 2, 113, 188
   and quadratic factor, 119
   and third-order system, 120
   construction, 114
boundary condition, 4
braking torque, 72
branches, 53

cascade control, 144, 222
characteristic equation, 75, 81
closed loop
   damping, 161
   frequency response, 192, 197
   stability, 154
   system, 70
   transfer function, 75
combined inputs, 6, 8
compensator, 204
   and inverse polar plot, 111
   design, 218
   design with Bode plot, 220
   polar plots, 106
complementary function, 4
complex roots, 75
compressibility, 39
computer package plots, 120
computer root locus plot, 163
conservation
   equation, 43
   of mass, momentum, energy, 22
   principle, 22

constant M and N circles, 194
continuity principle, 22
continuous-flow stirred-tank, 42
continuous stirred-tank reactor, 46
controller gain, 75
convolution integral, 9, 13, 14
cover-up rule, 10
current source, 25

d'Alembert forces, 33, 40
damping coefficient, 84
delay, 5
derivative action, 2, 130
   and closed loop poles, 131
deviations from steady-state, 34
differential equations, 2
distributed parameter system
   model, 47
distributed variable, 40
disturbance
   input, 141
   rejection, 204

electrical systems, 24, 97
electromechanical systems, 37
equilibrium condition, 17

feedback, 2, 70
   control, 70
feedforward, 2
   control, 70, 141, 222
final value theorem, 12
first-order
   phasor diagram, 96
   polar plot, 102
   signal flow graph, 55
   system, 3
flow graph from functional
   diagram, 57
fluid
   drive, 41
   mixers, 43
fourth-order system, 12

frequency
   domain, 2, 173
   response, 93
   transfer function, 93

gain margin, 181, 185, 188, 206
general solution, 4, 5

heat flow and analogue, 45
hydraulic system, 39, 88
   with relief valve, 92

impulse
   input, 3
   response function, 13
independent variable, 1
inertia with damping, 35
initial condition, 3, 5, 10, 15, 28
input variable, 15
integral action, 2, 128, 130
   time, 128
interaction, 223
intermediate variables, 38
inverse
   Nyquist plot, 184
   polar plot, 107
   and inner loop, 112

Kirchhoff's laws, 24

lag-lead compensator, 206
Laplace transform, 1, 6
   inversion, 10
   pairs, 7
   table of, 7
   theorems, table of, 9
linear system, 3
   modelling, 22
linear time invariant continous
   system, 3
linearization, 16, 40
logarithmic plots, 113
loop gain and closed loop poles,
   84

loop interaction, 222

magnitude ratio, 93
marginally stable systems, 82
Mason's rule, 54, 56
mass—damper system, 29
mass—spring—damper system, 32, 98
mathematical modelling, 1
Matlab plots, 121
measurement noise, 218
mechanical and electrical analogues, 36
mechanical systems, 29
mixed tanks in series, 44
mixer, 41
   control, 143, 147
modelling, 1

negative feedback loop, 59, 70
nested flow graph, 57
Nichols chart, 194, 196
nodes, 53
noise, 131
non-interacting controllers, 224
Nyquist
   criterion, 173
   path, 173
   plots, 2, 101, 174, 182

open loop
   control, 70
   pole and closed loop behaviour, 86
   zero and closed loop behaviour, 87
output
   disturbance, 218
   variable, 15

parallel circuit, 26
parameter variation, 85
partial fractions, 10
particular integral, 4
phase
   advance compensator, 206, 212
   angle, 93
   compensators, 2, 204
   lag compensator, 206, 209, 214
   lead compensator, 206, 212
   margin, 181, 185, 188, 206, 212
phasor diagram, 2, 94
PID control, 132, 217
polar plots, 101
pole
   polynomial, 81
   positions, 82

pole—zero
   combination, 116
   transfer function, 204
poles, 2, 81
position control, 138
process
   interaction, 223
   system analogue, 41
   systems, 40
proportional control, 70
proportional-derivative
   control, 133, 137, 164, 178
   controller, 215
proportional-integral
   action, 164, 178
   control, 133
   controller, 215
proportional-integral-derivative
   control, 132, 217
pulse input, 4, 23
pure delay, 182
   and Bode plot, 190
   and polar plot, 105

quadratic factor, 12

ramp input, 10, 16, 24
RC circuits, 24
RCL circuit, 26, 28
relative stability, 181
reset action, 128
right-half plane zero, 187
RL circuits, 25
robot controller, 144
root locus,
   construction, 160
   diagram, 158
   for lead compensator, 208
   plot, 2, 154, 158
   with lag compensator, 208
rotary inertia system, 33, 74
Routh array, 156
Routh—Hurwitz criterion, 2, 154, 168

saw-tooth input, 24
second-order
   phasor diagram, 96
   polar plot, 102
   signal flow graph, 55
   system, 4, 10
   transfer function, 15
series circuit, 26
servomechanism, 37
signal flow graph, 2, 53
   algebra, 53
   for cascade controller, 148
   reduction, 54

simple feedback, 70
simple pole, 109
single-input single-output system, 2
sinusoidal inplut, 2
solids mixer, 52
speed control, 138
spool valve, 39
spring—damper system, 31, 99
stability, 2, 81, 82, 154, 173
   and the Bode plot, 188
   margin, 86
stable/unstable poles, 82
standard inputs, 23
steady-state sinusoidal response, 95
step input, 4
stiction, 74
superposition, 3
   principle, 6
suspension system, 99
swing door, 35
system
   analogue, 22
   poles, 81
systems in series, 5

tank flow
   model, 3
   problem, 8
third-order polar plot, 102, 104
time constant, 3, 25, 26, 37
time invariant system, 3
transfer function, 2, 7, 15, 53, 100
   from signal flow graph, 76

undamped natural frequency, 84
unit
   impulse, 13
   ramp input, 13
   step, 4, 6
unstable
   poles, 180
   system, 186

velocity feedback, 2, 76, 136
viscous
   coupling, 33
   damping, 71

weighting function, 13

zero polynomial, 81
zeros, 2, 81, 177
Ziegler and Nichols controllers, 134